CHEMICAL TECHNOLOGY

Key Developments in Applied Chemistry,
Biochemistry and Materials Science

CHEMICAL TECHNOLOGY

Key Developments in Applied Chemistry, Biochemistry and Materials Science

Edited by

Nekane Guarrotxena, PhD,
Gennady E. Zaikov, DSc, and A. K. Haghi, PhD

AAP | APPLE ACADEMIC PRESS

Apple Academic Press Inc.
3333 Mistwell Crescent
Oakville, ON L6L 0A2
Canada

Apple Academic Press Inc.
9 Spinnaker Way
Waretown, NJ 08758
USA

Exclusive worldwide distribution by CRC Press, a member of Taylor & Francis Group

Printed in the United States of America on acid-free paper

International Standard Book Number-13: 978-1-77188-051-0 (Hardcover)

Library and Archives Canada Cataloguing in Publication

Chemical technology (Apple Academic Press)
Chemical technology: key developments in applied chemistry, biochemistry and materials science / edited by Nekane Guarrotxena, PhD, Gennady E. Zaikov, DSc, and A. K. Haghi, PhD.

Includes bibliographical references and index.
ISBN 978-1-77188-051-0 (bound)
1. Chemistry, Technical. 2. Biochemistry. 3. Materials science. I. Guarrotxena, Nekane, editor II. Zaikov, G. E. (Gennadii Efremovich), 1935-, author, editor III. Haghi, A. K., author, editor IV. Title.

TP145.C54 2015 660 C2015-900408-X

Library of Congress Cataloging-in-Publication Data

Chemical technology (Apple Academic Press)
Chemical technology : key developments in applied chemistry and materials science / [edited by] Nekane Guarrotxena, PhD, Gennady E. Zaikov, DSc, A.K. Haghi, PhD.

pages cm
Includes bibliographical references and index.
ISBN 978-1-77188-051-0 (hardback)
1. Chemistry, Technical. 2. Biochemistry. 3. Materials science. I. Guarrotxena, Nekane, editor. II. Zaikov, G. E. (Gennadii Efremovich), 1935- editor. III. Haghi, A. K., editor. IV. Apple Academic Press. V. Title.

TP145.C43 2015 660--dc23 2015000865

Apple Academic Press also publishes its books in a variety of electronic formats. Some content that appears in print may not be available in electronic format. For information about Apple Academic Press products, visit our website at **www.appleacademicpress.com** and the CRC Press website at **www.crcpress.com**

CONTENTS

LIST OF CONTRIBUTORS

O. M. Alekseeva
Emanuel Institute of Biochemical Physics, Russian Academy of Sciences, Moscow 119334, Email: olgavek@yandex.ru

Sogrina Darya Alexandrovna
Moscow State University of Food Production, 11 Volokolamskoe Shauosse, Moscow 125080, Russia

D. S. Andreev
Volgograd Architectural University Sebryakovsky Branch

M. Arabani
Professor, Faculty of Engineering, University of Guilan, Rasht 3756, Iran, Email: arabani@guilan.ac.ir

V. A., Babkin
Volgograd Architectural University Sebryakovsky Branch, Email: Babkin_v.a@mail.ru

Bakr Mona
The National Institute for Laser Enhanced Sciences, Cairo University, Egypt

A. A. Belov
University of Chemical Technology of Russia (RChTU) him. D. I. Mendeleev, ch. Biotechnologies, Research Institute of Textile Materials, Moscow, Russian Federation, Email: ABelov2004@ yandex.ru

Samarth Bhatt
Jena University Hospital, Friedrich Schiller University, Institute of Human Genetics, Kollegiengasse 10, D-07743 Jena, Germany.

V. I. Binyukov
N. M. Emanuel Institute of Biochemical Physics, Russian Academy of Sciences, ul. Kosygina 4, 119334 Moscow, Russian Federation. Fax (7-495) 137 41 01, tel. (7-495) 939 71 40

S. B. Bokieva
M. V. Lomonosov MSU, Biological Faculty, Leninskie Gory, 119991 Moscow, Russia

S. N. Bondarenko
Volzhsky Polytechnical Institute, branch of Federal State Budgetary Educational Institution of Higher Professional Education, Volgograd State Technical University, 42a Engels Str., 404121, Volzhsky, Volgograd Region, Russia

E. B. Burlakova
Emanuel Institute of Biochemical Physics RAS, 119334, Kosygina str., 4, Moscow, Russia

N. I. Chekunaev
Semenov Institute of Chemical Physics, Russian Academy of Sciences (RAS), Ul. Kosygina, 4, Moscow, 119991, Russia. nichek@mail.ru

M. S. Chirikova
Institute of Microbiology, National Academy of Sciences, Belarus, 220141, Kuprevich str.2, Minsk, Belarus, E-mail: margarita.chirikova@mail.ru, fax: +375(17) 267-47-66

Tarek A. El-Tayeb
The National Institute for Laser Enhanced Sciences, Cairo University, Egypt

Vladimir S. Feofanov,
N. M. Emanuel Institute of Biochemical Physics of the Russian Academy of Sciencesv, Kosygin st. 4, 117977 Moscow, Russia

Sergey N. Gaydamaka
Department Chemical Enzymology, Chemistry Faculty, Moscow State University, Leninskye Gory, 1, build.11, Moscow,119992, Russia, Phone: +7(495) 939-5083, Fax: + 7 (495) 939-5417, e-mail: s.gaidamaka@gmail.com

Sergey Gaydamaka
Murygina Lomonosov Moscow State University, Chemistry Faculty, Department of Chemical Enzymology. 119991, Moscow, Leninsky gory 1/11, fax: +7-495-939-54-17., e-mail: s.gaidamaka@gmail.com

N. Yu. Gerasimov
Emanuel Institute of Biochemical Physics RAS, 119334, Kosygina str., 4, Moscow, Russia, e-mail: n.yu.gerasimov@gmail.com

M. D. Goldfein
Saratov State University named after N.G. Chernyshevsky, Russia, goldfeinmd@mail.ru
A. N. Goloshchapov
Emanuel Institute of Biochemical Physics RAS, 119334, Kosygina str., 4, Moscow, Russia

Iman E. Gomaa
German University in Cairo (GUC), Main Entrance of Al-Tagamoa Al-Khames, New Cairo City, P.O. 11835, Egypt, German University in Cairo, Egypt, iman.gomaa@guc.edu.eg

N. A. Grebenkina
Higher Chemical College, RAS, Miusskaya sq., 9 Moscow, 125047

K. Z. Gumargalieva
N. N. Semenov Institute of Chemical Physics, RAS, 4 Kosygin Street, Moscow, 119334, Russia

A. K. Haghi
Faculty of Engineering, University of Guilan, Rasht, Postal code: 3756, I. R. Iran. Tel: +98(131)6690270, Fax: +98 (131) 6690270. Email:Haghi@guilan.ac.ir

A. N. Inozemtsev
M. V. Lomonosov MSU, Biological Faculty, Leninskie Gory, 119991 Moscow, Russia

V. F. Kablov
Volzhsky Polytechnical Institute (branch) Volgograd State Technical University, 42a Engelsa Street, Volzhsky, Volgograd Region, 404121, Russian Federation, E-mail: vtp@volpi.ru; www.volpi.ru

A. M. Kaplan
Semenov Institute of Chemical Physics, Russian Academy of Sciences (RAS), Ul. Kosygina, 4, Moscow, 119991, Russia.

O. V. Karpukhina
N. N. Semenov Institute of Chemical Physics, RAS, 4 Kosygin Street, E-mail: olgakarp@newmail.ru

V. V. Kasparov
Emanuel Institute of Biochemical Physics RAS, 119334, Kosygina str., 4, Moscow, Russia

N. A. Keibal
Volzhsky Polytechnical Institute (branch) Volgograd State Technical University, 42a Engelsa Street, Volzhsky, Volgograd Region, 404121, Russian Federation, E-mail: vtp@volpi.ru ; www.volpi.ru

Yu. A. Kim
Institute of Cell Biophysics, Russian Academy of Sciences, Pushchino, Moscow region

Lubov Kh Komissarova
N. M. Emanuel Institute of Biochemical Physics of the Russian Academy of Sciences, Kosygin st. 4 117977 Moscow, Russia. e-mail: chembio@sky.chph.ras.ru, e-mail komissarova–lkh@mail.ru, Telephones: 8(495)9361745(office), 8(906)7544974(mobile), Fax: (495)1374101

A. I. Korotaeva
University of Chemical Technology of Russia (RChTU) him. D.I.Mendeleev, ch. Biotechnologies, Research Institute of Textile Materials, Moscow, Russian Federation

A. L., Kovarskij
Emanuel Institute of Biochemical Physics RAS, 119334, Kosygina str., 4, Moscow, Russia

T. V. Krekaleva
Volzhsky Polytechnical Institute (branch) Volgograd State Technical University, 42a Engelsa Street, Volzhsky, Volgograd Region, 404121, Russian Federation, E-mail: vtp@volpi.ru ; www.volpi.ru

Thomas Liehr
Jena University Hospital, Friedrich Schiller University, Institute of Human Genetics, Kollegiengasse 10, D-07743 Jena, Germany

E. I., Martirosova
Emanuel Institute of Biochemical Physics, RAS Kosygina st., 4 Moscow, 119334, e-mail: ms_martins@mail.ru

L. I. Matienko
N. M. Emanuel Institute of Biochemical Physics, Russian Academy of Sciences, ul. Kosygina 4, 119334 Moscow, Russian Federation. Fax (7-495) 137 41 01, tel. (7-495) 939 71 40, e-mail: matienko@sky.chph.ras.ru

L. A. Mosolova
N. M. Emanuel Institute of Biochemical Physics, Russian Academy of Sciences, ul. Kosygina 4, 119334 Moscow, Russian Federation. Fax (7-495) 137 41 01, tel. (7-495) 939 71 40,

Valentina P. Murygina
Department Chemical Enzymology, Chemistry Faculty, Moscow State University, Leninskye Gory, 1, build.11, Moscow,119992, Russia, Phone: +7(495) 939-5083, Fax: + 7 (495) 939-5417, e-mail: vp_murygina@mail.ru, vpm@enzyme.chem.msu.ru

O. V. Nevrova
Emanuel Institute of Biochemical Physics RAS, 119334, Kosygina str., 4, Moscow, Russia

B. K. Novosadov
V. I. Vernadsky Institute of Geochemistry and Analytical Chemistry RAS, 119991 Moscow, Kosygina str., 19, Russian Federation, e-mail: bk.novosadov@mail.ru

I. G. Plashchina
Emanuel Institute of Biochemical Physics, RAS Kosygina st., 4 Moscow, 119334

D. A. Provotorova
Volzhsky Polytechnical Institute, branch of Federal State Budgetary Educational Institution of Higher Professional Education Volgograd State Technical University, 42a Engels Str., 404121, Volzhsky, Volgograd Region, Russia

E. A. Raspopova
University of Chemical Technology of Russia (RChTU) him. D.I.Mendeleev, ch. Biotechnologies, Research Institute of Textile Materials, Moscow, Russian Federation

G. K. Rossieva
Volgograd Architectural University Sebryakovsky branch

E. G. Rozantsev
Saratov State University named after N.G. Chernyshevsky, Russia

A. S. Samsonova
Institute of Microbiology, National Academy of Sciences, Belarus, 220141, Kuprevich str.2, Minsk, Belarus

T. P. Shakun
Institute of Microbiology, National Academy of Sciences, Belarus, 220141, Kuprevich str.2, Minsk, Belarus

A. G. Stepanova
Volzhsky Polytechnical Institute (branch) Volgograd State Technical University, 42a Engelsa Street, Volzhsky, Volgograd Region, 404121, Russian Federation, E-mail: vtp@volpi.ru ; www.volpi.ru

R. Tanzadeh
Department of Civil Engineering.University of Guilan, Rasht, Iran.Tel: +98(131)3229883, Fax: +98 (131) 3231116. Email: rashidtanzadeh@yahoo.com

P. Valentina
Murygina Lomonosov Moscow State University, Chemistry Faculty, Department of Chemical Enzymology. 119991, Moscow, Leninsky gory 1/11, fax: +7-495-939-54-17.

Ananiev Vladimir Vladimirovich
Moscow State University of Food Production, 11 Volokolamskoe shauosse, 125080, Moscow, Russia, kaf.vms@rambler.ru

G.E. Zaikov
N. M. Emanuel Institute of Biochemical Physics, Russian Academy of Sciences, ul. Kosygina 4, 119334 Moscow, Russian Federation. Fax (7-495) 13741 01, tel. (7-495) 939 71 40,-mail: chembio@sky.chph.ras.ru

A. Zhivaev
Volzhsky Polytechnical Institute (branch) Volgograd State Technical University, 42a Engelsa Street, Volzhsky, Volgograd Region, 404121, Russian Federation, E-mail: vtp@volpi.ru ; www.volpi.ru

LIST OF ABBREVIATIONS

AP	acetophenone
AHB	alkylhydroxybenzenes
AR	alkylresorcinols
AC	asphalt concrete
AFM	atomic force microscopy
BNCT	boron neutron capture of tumor therapy
BSA	bovine serum albumin
CNTs	carbon nanotubes
COD	chemical oxygen demand
CNR	chlorinated natural rubber
CMC	critical micelle concentration
DAAD	deutsche akademische austausch dienst
DAC	dialdehydecellulose
DP	diamond pore
DSC	differential scanning calorimeter
DNSA	dinitrosalicylic acid
DSB	double strand breaks
DM	dressing materials
EMEM	eagle's minimal essential medium
EPR	electron paramagnetic resonance
ETP	electron transport particles
ERKs	extracellular signal-regulated kinases
GC	gas chromatograph
GCMD	grand canonical molecular dynamics
GCMC	grand canonical monte carlo
HMPA	hexamethylphosphorotriamide
HR	hexylresorcinol
HMS	high melt strength
HCO	hydrocarbon oxidizing cells
ISCN	international system for human cytogenetic nomenclature
IUPAC	international union of pure and applied chemistry

LED	light emitting diode
LMWC	low molecular weight chitosans
MSD	mean-square displacement
MFI	melt flow index
MPC	methylphenylcarbinol
MR	methylresorcinol
MF	microfiltration
MD	molecular dynamics
MWCO	molecular weight cut-off
MC	monte carlo
MWNT	multi-walled carbon nanotube
NF	nanofiltration
NILES	National Institute of Laser Enhanced Sciences
NCT	neutron capture therapy
NDT	nottingham device test
PEH	phenyl ethyl hydroperoxide
PB	phosphate buffer
PBS	phosphate buffer saline
PTT	photo thermal therapy
PCN	polymer–clay nanocomposites
PP	polypropylene
RESPA	reference system propagator algorithm
RO	reverse osmosis
SEM	scanning electron microscope
SSB	single strand breaks
SWNTs	single-walled carbon nanotubes
SEM	standard error of the mean
SP	straight path
SBS	styrene–butadiene–styrene
TMC	thermomechanical curves
TFOT	thin film oven test
TEM	transmission electron microscopic
UF	ultrafiltration
VACF	velocity autocorrelation function
ZP	zigzag path

LIST OF SYMBOLS

ρ	material density
D_S^*	preexponential factor
A	difference between titration results in test and control samples
A	new generated area at crack penetration
a_S	crack radius
B	concentration of enzyme solution sample
C and n	material parameters
E	modulus of elasticity
L	membrane thickness
LC	lipolytic activity
m	electron rest mass
$m1$	mass of weighing bottle
$m2$	net bottle weight
T	alkali titer
V	volume of analyzed sample
$V1$	extract aliquot volume
$V2$	volume of graduated flask
v_m	limiting crack velocity
$x1$	total concentration of substances

PREFACE

This collection presents to the reader a broad spectrum of chapters in the various branches of industrial chemistry, biochemistry, and materials science which demonstrate key developments in these rapidly changing fields.

This book offers a valuable overview and myriad details on current chemical processes, products, and practices. The book serves a spectrum of individuals, from those who are directly involved in the chemical industry to others in related industries and activities. It provides not only the underlying science and technology for important industry sectors, but also broad coverage of critical supporting topics.

This new book:

- is a collection of chapters that highlight some important areas of current interest in industrial chemistry, biochemistry, and materials science
- focuses on topics with more advanced methods
- emphasizes precise mathematical development and actual experimental details
- analyzes theories to formulate and prove the physicochemical principles
- provides an up-to-date and thorough exposition of the present state-of-the-art complex materials
- familiarizes the reader with new aspects of the techniques used in the examination of polymers, including chemical, physicochemical, and purely physical methods of examination
- describes the types of techniques now available to the chemist and technician and discusses their capabilities, limitations, and applications

CHAPTER 1

THE USE OF ULTRASOUND FOR FOAMING OF POLYPROPYLENE

ANANIEV VLADIMIR VLADIMIROVICH and
SOGRINA DARYA ALEXANDROVNA

CONTENTS

1.1 INTRODUCTION

Wide application of polymeric foamed materials, based on the polyolefin's, explained by their mechanical, insulating, and operational properties. Foamed polypropylene (PP) is widely used material in numerous applications. Foamed PP products are highly demanded in automobile production, construction; some of them are used as a cushioning materials and weight reducers in complex structures. There are some practically used manufacturing technologies to produce foamed PP. But all of them have some general disadvantages. All of them are very complex and, hence, this technologies are expensive, making PP foam more costly them other foamed polyolefins. Frequently, foaming of PP requires the use of specific blowing agents, specific types of polymer, and special process conditions [1].

Analysis of the literature data has shown that significant impact on technological and operational properties of the polymer products renders ultrasonic treatment during their production [2–4]. With the changes occurring in polymers under sonication, we assume that using of ultrasonic irradiation may be promising for the foaming processes. Moreover, we presuppose that ultrasonic treatment during extrusion of polymer foam can change structure (size of cells, cells distribution in the material volume) of foamed PP and by that change bulk density of foamed materials.

1.2 EXPERIMENTAL PART

Russian PP grade 21020, azodicarbonamide and Hydrocerol BM 70, as blowing agent, have been chosen for foaming. Usually, specially designed PP grades are used for foaming processes. All of them have high branching and long side chains of their macromolecules. It is considered that such polymer structure provides high melt strength (HMS) of this PP grades (over six times higher than traditional PP grades). We thought it best to use Russian PP grades, to illustrate the benefits of the use of ultrasound in the foaming process. Therefore, we chose a PP grade with the same rheological characteristics as special grades with HMS (melt flow index of used PP = 2 g/10 min).

The blowing agent, Hydrocerol BM 70, is injected in amounts from 0.1 wt % to 1.5 wt % of the composition. According to current technology chemical blowing agents on the basic of azodicarbonamides injects into composition in quantity 0.5 wt %–1 wt %, and 5 wt %. We assumed that the use of ultrasound in the extrusion process can improve the gassing in the polymer matrix. So, we decided to decrease the concentration of the blowing agent to 0.1 wt %.

Prerequisite for such solutions were previously conducted experiments in which we observed a significant reduction in the melt viscosity of PP, which was measured directly during the extrusion process. It is suggested in literature date, that significant pressure changes may occur in these areas. Moreover, a significant increase in temperature can take place in the zones of small volume. We assumed that these zones, arising from the inhomogeneity of the melt, may play a role of nucleators. The sharp reduction in pressure in these areas will promote growth of gas bubbles.

Selection of blowing agent explains that it is easy to blend them with polymer, and also temperature decomposition range of azodicarbonamide corresponds to the processing temperature of PP. Azodicarbonamide decomposes with allocation of of nitrogen at the temperature about 170°C.

Hydrocerol BM 70 is a chemical substance that decomposes or reacts by the influence of heat. This is chemical, endothermic foaming agent. To achieve an optimum gas yield, a processing temperature of at least 180°C is suggested.

At the present time double-screw extruder and single-screw extruders with specific construction of screw are used for production of foamed materials. Double-screw extruders provide a better homogenization of the polymer melt then other types of equipment. They provide a more uniform distribution of blowing agent and nucleator in the volume of polymer during foaming processes. Based on the analysis of modern equipment for PP foaming, it was decided to create two special laboratory installations.

One of them is made on the basis of the double-screw extruder, with screws diameter 20 mm. It is used for obtaining of stands and cylindrical pellets. An installation includes pressure sensor, unit of ultrasonic processing, extrusion die (for one or two stands formation), bath for stands cooling, pulling device, and granulator. Azodicarbonamide is used as a blowing agent. Maximum productivity of installation is 8 kg/h.

Unit of ultrasonic processing is a bundle of ultrasonic vibration generator (oscillation frequency 22 kHz and capacity 1.5 kW), magnetostrictive vibration transducer, titanium sonotrode, and melt treatment chamber, which is attached to the outlet flange of the extruder. At the output of the camera is set strand die.

The second installation is made using the single-screw extruder, with barrier screw, screw diameter 12 mm. Hydrocerol BM-70 is used as a blowing agent. Maximum productivity of installation is 1.5 kg/h. An installation includes: pressure sensor, unit of ultrasonic processing and a set of capillaries for rheological studies directly during extrusion. The kit of ultrasonic unit includes an ultrasonic generator (oscillation frequency 22 kHz and capacity 300 W), piezoceramic vibration transducer, and titanium sonotrode. At the output of the camera is set slit die (width 100 mm). A system of rolls with air cooling is used for receiving of flat film. Thus, on the installation with a single-screw extruder, a strand can be obtained as well as the film material. Appearance of installations is shown in Figure (1.1a, b).

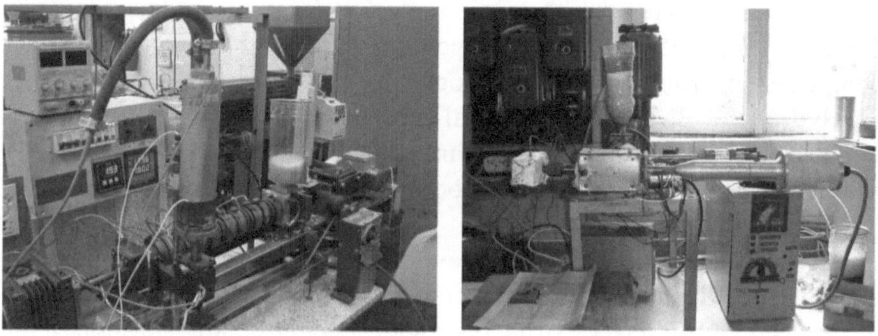

(a) (b)

FIGURE 1.1 Appearance of laboratory extrusion systems based on: (a) Double-screw extruder and (b) single-screw extruder.

The sonication of the polymer melt was carried out directly in the extruder under laboratory conditions by using apparatus for ultrasonic treatment. Without ultrasonic, ultrasound block was not dismantled, to provide constancy of process conditions.

Ultrasonic apparatus for processing the polymer melt represents a complicated system of blocks and elements consisting of transducer of electrical oscillations, system for concentrating of ultrasonic vibrations, instrument for input of ultrasonic vibrations, ultrasonic vibration generator, system of control, and automation.

Beforehand, it was experimentally established that neither type of generator or electroacoustic transducer does not affect the results obtained in the process modification polymer melt.

Order of the experiment:

(i) Preparing the mixture of polymer and blowing agent: after weighing the initial components, they were subjected mixing on a mechanical shaker; optimum mixing time was established experimentally, it is a time, when the best quality of mixing is achieved;
(ii) Loading the mixture into the hopper of the extruder;
(iii) Setting process parameters, such as rotation frequency of the screw; temperatures by processing zones of extruder; the pressure in the extruder head was constantly controlled during foaming, it has allowed to estimate the rheological properties of polymer melt;
(iv) Heating of the extruder: heating time has been established empirically, for this purpose was evaluated melt quality of initial PP by the absence of not melted areas;
(v) Obtaining of the samples: samples were prepared in the form of strands, or in the form of films, and complete set of installations and extrusion mode was same for samples, obtained with and without ultrasonic treatment.

It should be noted that PP foaming without ultrasonic treatment was accompanied by abrupt changes in pressure (both strands and films), although abrupt changes in pressure were not observed during foaming with ultrasound. PP samples foamed without ultrasonic treatment had such defects as the surface roughness and polythickness. Samples foamed with ultrasonic treatment of its melt had even surface and uniform thickness over the length. It should be mentioned that formation of different thickness strands had random nature, although the occurrence of irregular portions of smaller diameter is associated with nonuniform nucleation in PP, foamed without ultrasound. Reducing defects in the treated samples indicates that ultrasound promotes more uniform nucleation and growth of the bubbles in the foam. In previously studies conducted by us, we found that ultrasonic treatment of the various polymer melts facilitates formation of more uniform and amorphized structure which shall give rise to favorable conditions for the foaming processes of the melt. Another reason also improving the conditions for foaming is a better distribution of the components introduced into the polymer melt under the action of ultrasonic vibrations. The result is a more uniform nucleation and growth of bubbles in the foamed material and reduction of defects in the processed samples.

Furthermore, we can assume that the formation of homogeneous structures promote cavitation phenomena occurring under the influence of ultrasound. Usually, cavitation is understood as fluid rupture under creating negative pressures therein. Under certain tension liquid is broken, forming a cavity in which the pair and dissolved gasses quickly penetrate. Low strength liquids to tensile stress associated with the presence in it of various irregularities, such as microbubbles, which are the cavitation nucleus. Currently in various literary sources [5] describes the following mechanism of cavitation phenomena, under the influence of sound vibrations compression cavitation bubbles was carried out, but there was practically no effect of the collapse. Instead, the effect of distortion of the spherical shape of the cavitation bubble, loss of stability, and splitting to form smaller bubbles which pulsate in sync with the acoustic field are observed. Microcavitation during foaming may contribute new pores and, consequently improve the foam structure.

Next, the samples were examined by the following methods:

(i) Method of determining the melt flow index (MFI) is used to estimate the rheological properties of PP. The test was conducted on the special device intended for laboratory determination of MFI for powdered, granulated, and pressed thermoplastics. It was established that PP has a melt flow index 2.26 g/min, which is as claimed by the manufacturer MFR value for used PP grade.

(ii) Tensile test method was used to determine the physicomechanical characteristics of the samples. We used the method of State Standard 14236–81 for testing. Polymer films tensile test methods, because there is no corresponding standard for the materials in the form of strands. We used the method of State Standard 14236–81 for testing of film materials too. The experiment was conducted on a tensile testing machine type RM-10. Because a heterogeneous material was used for test samples, it was important to obtain accurate data at any given time. Thus, breaking machine with a computer interface was used for testing, allowing to adjust the load up to 10,000 measurements per second. Limit of allowable measurement error in the direct course does not exceed ±1 percent of the measured load.

(iii) Method of constructing thermomechanical curves by determining the dependence of deformation on the temperature. Samples in the form of thread are changed for testing. Experiment was carried out on a testing machine TRM 251. With a special computer program parameters setting, such as heating time, heating temperature, and holding time, that is, the time during which the preset temperature fixed, were conducted. Registration of thermomechanical curves and achieved values of elongation as well performed automatically which resulted in high measurement accuracy and to establish the intervals transition temperature of the sample from one physical state to another.

(iv) Method of optical microscopy. Tests were conducted using an optical microscope Polam equipped with a digital camera Nikon, recording the obtained images in electronic form. Microscopic examination was carried out as in transmitted and reflected light.

1.3 RESULTS AND DISCUSSION

Results of the tensile testing of samples in the form of strands obtained in the extrusion temperature 210°C and a blowing agent content of 1 percent in the composition shown in Table 1.1.

TABLE 1.1 Physicomechanical properties of foamed PP

Tested Material	Breakdown Voltage at Yield (MPa)	Percent Elongation with Rupture (%)
PP samples, obtained by extrusion without treatment	12.2 ± 1	18 ± 2
PP samples, obtained by extrusion with ultrasound treatment	14.3 ± 1	35 ± 2

It should be noted that ultrasonic treatment causes an increase of elongation approximately two times. Although increase in breakdown voltage at yield is observed under the influence of ultrasound, the obtained difference values for treated and untreated samples exceeds the experimental error. However, it must be taken into account that the degrees of foaming of samples obtained by sonication were significantly higher than without the modification. It was seen visually, and then confirmed by light microscopy. Because a measurable voltage break of the sample is calculated from the value of the initial cross-section of samples, in modified objects the real value of cross section due to the polymer was significantly less than that of the unmodified and true stress in the polymer section they have, respectively, more. Therefore, we can conclude that ultrasound sonication of the melt during foaming leads to improvement of strength as well as the deformation characteristics of the polymer matrix.

A method of constructing of thermomechanical curves (TMC) as well as the stretching method is sensitive to a change in true cross section of the polymer matrix, because deformation of the sample occurs under constant load. According to the analysis of TMC, it is possible not only to judge the temperature transition intervals of polymer from one physical state to another, but also indirectly reveal the presence of a difference in shaping the structure of the samples treated with ultrasound and without treatment. Value of the error in constructing of TMC was 5 percent. TMC of foamed PP are shown in Figure 1.2.

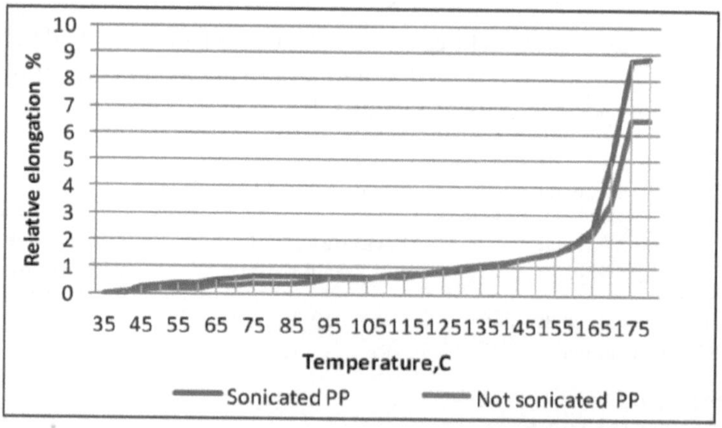

FIGURE 1.2 Thermomechanical curves of foamed PP, obtained at a temperature of 210°C and a modifier content of 1 percent. Blue line: Sample, produced without exposure to ultrasound. Red: Samples foamed with ultrasound.

A decrease of glass transition temperature and increase in elongation PP, foamed with ultrasound indicates a change in the structure of samples. Thus, for the sample treated with ultrasound, an increase resistance to temperature is characteristic.

Date of TMC and physicomechanical investigations shows an improvement in strength characteristics of irradiated PP samples. Such improvement can be caused by the formation of more homogeneous structure, foam cells grows, and reducing the number of defects.

To confirm the assumption that ultrasonic treatment caused changes in the structure the foam, we examined the samples by optical microscopy method. Figure 1.3 shows photomicrographs of foamed samples, prepared under the same conditions and with the same magnification. Given photographs were obtained in transmitted light. They show the general nature of pore distribution in the samples.

Analyzing the micrographs of foamed PP, it is easy to see that for the same parameters of the extrusion process, sonication leads to increasing of the number gas cells and an increasing in their average size. Increasing extrusion temperature at 10°C also leads to an increase in the number of pores and increase their sizes. The effect from action of ultrasound is stored at elevated temperatures, but sonication promotes the formation of a homogeneous structure with a uniform distribution of cells by volume of the polymer matrix. We plan to get more complete information spending microscopic examination of cross sections of the samples.

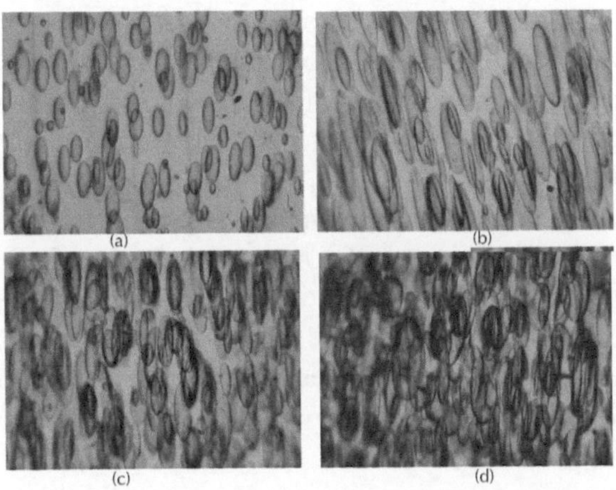

FIGURE 1.3 Optical microphotos of foamed polymeric PP samples: (a) PP sample produced at a temperature of 210°C and containing 0.6 percent of a blowing agent without the influence of ultrasound, (b) PP sample produced at a temperature of 210°C and containing 0.6 percent of the blowing agent under the influence ultrasonic, (c) PP sample produced at a temperature of 220°C and containing 0.6 percent of a blowing agent without the influence of ultrasound, and (d) PP sample produced at 220°C and containing 0.6 percent of the blowing agent under the influence ultrasound.

Further it is planned to conduct a more detailed study of foamed PP structure by electronic microscopy method, study the effect of the concentration of the blowing agent in the foam properties, and determine the effect of foaming conditions on the quality of the final products.

1.4 CONCLUSION

(i) According to the researches carried out, it can be concluded that ultrasonic treatment of the PP melt improves the basic mechanical characteristics of foamed PP. Analysis of TMC shows that ultrasonic treatment during helps to increase elongation. Therefore, we can assume that ultrasonic vibration contribute to a more uniform distribution of the pores in volume of the material. This results in an increase of strength characteristics of the material when tested in one axial tension mode, as well as a slight increase in elongation of the samples under study by method of construction of TMC.

(ii) On the results of research on the structure of foamed PP samples by optical microscopy can be concluded that sonicated PP samples have a more homogeneous structure. Thus, we can conclude that there is a theoretical possibility of changing the structure and properties of foamed materials based on PP by using ultrasonic treatment of their melts during the foaming process.

(iii) The use of ultrasonic treatment of the melt leads to a significant increase in the number of pores and their average size at the same extrusion conditions. Such events observed at different temperatures. The use of ultrasound allows the formation of a more uniform structure of the samples and reducing the number of defects, which is reflected in improved physical and mechanical properties of these samples.

KEYWORDS

- **Azodicarbonamide**
- **Blowing agent**
- **Foaming**
- **Microscopy**
- **Polypropylene**
- **Thermomechanical curve**
- **Ultrasound**

REFERENCES

1. Klempner, D.; and Sendzharevich, B; Polymeric Foams and Foam Technology. Profession: St. Petersburg; **2009**, 600 pp.
2. Ananev, V. V.; Gubanova, M. I.; Kirsh, I. A.; Semenov, G. V.; and Kozmin, D. V.; Modification of polyethylene initiated by ultrasound. Plastics, 6 (in Russian): Moscow, **2008**, 6–9 pp.
3. Semenov, G. V.; Ananev, V.V.; Kirsch, I. A.; Kozmin, D. V.; Gubanova. M. I.; Recycling of plastic waste under the influence of ultrasound. Plastics, 10 (in Russian): Moscow, **2008**, 41–44 pp.
4. Ananev, V. V.; Filinskaya, Y. A.; Kirsh, I. A.; Bannikova, O. A.; and Utkin, A. O.; Improving the quality of polymer composite materials and packaging design. Food Processing, 1 (in Russian): Moscow, **2012**, 16–18 pp.
5. Mason, T.; Translated from English. In: Kirkovski, L. I.; Ed. Chemicals and Ultrasound. Wiley: Lindley, J.; Davidson, R.; **1993**, 190 p.

CHAPTER 2

AG AND CO/AG NANOPARTICLES CYTOTOXICITY AND GENOTOXICITY STUDY ON HEP-2 AND BLOOD LYMPHOCYTES CELLS

IMAN E. GOMAA, SAMARTH BHATT, THOMAS LIEHR, MONA BAKR, and TAREK A. EL-TAYEB

CONTENTS

2.1 INTRODUCTION

Hyperthermia is a state where cells absorb more heat than they can dissipate, which can be lethal to the cells. Such phenomenon has been found as a promising approach for cancer therapy, because it directly kills cancer cells and indirectly activates anticancer immunity [1]. Because focusing the heat on an intended region without damaging the healthy tissue was an important problem that arose by the treatment with hyperthermia, targeting of a specific region was important. If nanoparticles can target a specific malignant tissue, hyperthermia can be directed to this specific tissue. Unlike the conventional hyper thermic techniques, some studies showed that using nanoparticles for hyperthermia helped cancer cells to reach the lethal temperature without damaging the surrounding tissue [2].

The choice of a specific light delivery mode in clinical settings is usually based on the nature and location of the disease. The optimal light dose can be obtained by adjusting the fluency rate and fluency element. The characterization of light penetration and distribution in solid tumors is important, because it will influence choosing a light source with an appropriate wavelength. From the light sources used in photo thermal therapy, the light emitting diode (LED) generating a desired high energy of specific wavelengths and can be assembled in a range of geometries and sizes [3].

Metal nanoparticles with their wide range of applications, such as catalytic systems with optimized selectivity and efficiency, optical components, targeted thermal agents for exploitation in drug delivery, and medical therapies, as well as surface-enhanced Raman spectral probing, have attracted research attention during the last decade. The size-dependent and shape-dependent properties of nanoparticles render them different from their corresponding bulk materials with macroscopic dimensions [4].

Because of their special physicochemical properties, metal nanoparticles showed a great progress in the bioanalytical and medical applications, such as multiplexed bioassays, biomedicine, ultrasensitive biodetection, [5] and bioimaging [6]. They have several biological applications in drug delivery, magnetic resonance imaging (MRI) enhancement, [6] and in cancer treatment [7].

Silver nanoparticles are among the noble metallic nanomaterials that have received considerable attention due to their attractive physicochemical properties. The surface plasmon resonance and the large effective scattering cross section of individual nanoparticles make them ideal candidates for biomedical applications [8]. On the other hand, other studies were directed toward synthesizing silver nanoshells of 40–50 nm outer surface diameter and 20–30 nm inner diameter using cobalt (Co) nanoparticles as sacrificial templates. In this case, the thermal reaction deriving force comes from the large reduction potential gap between the Ag^{+}/Ag and the Co^{+2}/Co redox couples which results in the consumption of Co cores and the formation of a hollow cavity of Ag nanoshells. The UV spectrum of this nanostructure exhibits a distinct difference from that of solid nanoparticles, which makes it a good candidate for application in photothermal materials [9].

As little is known about their biological applications in cancer treatment, and relying on the fact of being novel candidates providing high thermal effect, this study was directed toward investigation of the photothermal cancer therapy using silver nanoparticles and cobalt core silver shell nanoparticles in HEp-2 laryngeal cell carcinoma *in vitro*, as well as the side genetic effects both on DNA and chromosomal levels.

2.2 MATERIALS AND METHODS

2.2.1 SYNTHESIS OF AGNPS AND CO/AGNPS

The Ag and Co/Ag nanoparticles were obtained from the Nanotechnology Lab, the National Institute of Laser Enhanced Sciences (NILES), Cairo University. Spectrophotometric analysis was done to determine the absorption wavelengths range of Ag and Co/Ag NPs. Transmission electron microscope was used to investigate the shape and size range of these nanoparticles before using for application in cancer hyperthermia.

2.2.2 CELL CULTURE CONDITIONS

An established human epidermoid cancer cells (HEp-2) isolated from the larynx was obtained from American Type Culture Collection (CCL-23, ATCC, Rockville, MD) have been used for application of the *in vitro* tumor viability assay. Cells were routinely maintained in Eagle's minimal essential medium (EMEM) supplemented with 10 percent FBS and 1 percent antibiotic solution (including 10,000 U penicillin and 10 mg streptomycin) at 37°C in a humidified atmosphere of 5 percent CO_2 in air Cells

(100 cells/mm^2) were seeded in 24 well plates and grown to 75 percent confluence level overnight.

2.2.3 *IN VITRO VIABILITY ASSAY*

To ensure highest cytotoxic levels of tumor cells, the optimum light doses and nanoparticles concentrations were first determined by application of series of dark and light control experiments. Three replicates of HEp-2 laryngeal carcinoma cells were seeded overnight in 24 wells culture plates. Cells were incubated with different concentrations of the drugs of interest (Ag or Co/Ag) NPs in 5 percent FBS-containing medium and incubated at the standard culture conditions. On the next day, they were subjected to different monochromatic light doses from light emitting diode (LED) of 460 nm and 200 mW, while being incubated in 2 percent FBS medium. The distance between the light source and cell line surface was adjusted to 1 cm^2 spot surface area. Duration of light exposure was calculated with the following general formula; time (s) = (number of Joules × well surface area (m^2))/irradiance of LED (W/m^2).

Cells were washed with phosphate buffer saline (PBS) to get rid of the excess unabsorbed drug, and incubated with 0.0075 percent neutral red viability assay solution (Sigma Aldrich, N7005-1G) diluted in 2 percent FBS medium for 3 h. Excess dye was washed off with 0.9 percent NaCl, and the stained viable cells were lysed in lysis buffer (1% absolute ethanol: 1% distilled water: 0.02% glacial acetic acid) for 20 min with continuous shaking. Spectrophotometric measurement of cellular viability was performed in a 96 wells plate. Absorbance of the red color released after lyses of viable cells was measured using a plate reader (Victor 3V-1420, Germany) at 572 nm where the absorbance values indicate cell viability.

2.2.4 *GENOTOXICITY TEST (COMET ASSAY)*

The LC50 of both AgNPs as well as Co/AgNPs mediated PTT were used to evaluate the genotoxicity in freshly isolated peripheral blood lymphocytes. Blood lymphocytes were separated as reported by Singh et al [10]. Lymphocytes layer seen as a buffy coat was aspirated and washed with PBS then incubated overnight in RPMI growth medium containing 0.001 percent phytohemagglutinin. Equal numbers of cells were seeded in 24 wells plate for testing each experimental condition (treated cells with either AgNPs or Co/AgNPs at 50 J/cm^2, dark control and light control). These were compared with cells treated with the strong genotoxic chemotherapeutic drug Bleomycin, as a positive control. Viability of cells was monitored at 70 percent with reference to the control cells using trypan blue stain to exclude the cytotoxic effects of either drug upon interpretation of the genotoxicity results [11]. Comet assay was applied according to literature [12]. During electrophoresis, the broken DNA moves toward the anode forming a Comet tail. The greater the extent of DNA damage, the longer the tail length. Finally, 150 cells per slide were randomly selected for quantification of genotoxic effect of the tested drugs as a percentage of DNA damage using (Comet Imager, metasystem, version 2.2, GmbH) software.

2.2.5 MOLECULAR CYTOGENETIC TEST; MULTIPLEX FLUORESCENCE IN SITU HYBRIDIZATION (MFISH)

Blood collected from a healthy volunteer was cultured for 72 h in RPMI medium containing 1 percent penicillin/streptomycin, 15 percent FBS, and 2 percent phytohaemagglutinin. Duplicate whole blood cultures were treated with the same LC50 conditions used for the comet assay, before being harvested. Both dark and light control experiments were also included, to detect the sole effect of either the drug or the monochromatic light on exerting chromosomal aberrations. *In vitro* human lymphocyte assay for evaluating the chromosomal anomalies caused by chemical agents was applied on metaphases of duplicate slides from each sample referring to literature [13].

Slides prepared from lymphocyte cultures treated with either of the two drugs of interest and containing metaphase chromosomes were used for investigation of chromosomal aberrations by M-FISH. The M-FISH probe was prepared at the Institute of Human Genetics in Jena, Germany. After hybridization of two slides from each duplicate culture, slide washes and signal detection steps were applied as described in literature [14, 15]. Counterstain and antifade solution was applied using 4′, 6-diamino-2-phenylindole (DAPI) solution (Invitrogen, H-1200). The slides were stored in the dark at 4°C for 15 min before fluorescent microscopic examination. A total of 50 metaphases were analyzed using a fluorescent microscope

(Axio Imager.Z1 mot; Zeiss) equipped with the appropriate filter sets to discriminate between the five fluorochromes and the DAPI counterstain using a XC77 CCD camera with on-chip integration (Sony, Vienna, Austria). Image capturing and processing were carried out using an ISIS imaging system (MetaSystems, Altlussheim, Germany). Chromosomal gaps or breaks were estimated in comparison with a normal karyotype.

A number of 100 metaphases of each sample were analyzed by standard karyotyping using ISIS software (Metasystems GmbH, Altlussheim, Germany, 0017) for detection of the presence of any chromosomal aberrations. Karyotypes nomenclature was determined according to the International System for human Cytogenetic Nomenclature (ISCN). Both numerical and structural chromosomal aberrations were recorded for either AgNPs or Co/AgNPs treated cells compared with the control nontreated ones. Finally, the mitotic index per 1,000 blast cells for each treated sample with either drug of interest in comparison to the (untreated) cells was also assessed.

2.2.6 STATISTICAL ANALYSIS

Statistical analysis was applied using the GraphPad Prism 5.0 software. Results were obtained from three independent experiments and data were expressed as the mean ± standard error of the mean (SEM), using one-way ANOVA Tukey's multiple comparison test, and 95 percent confidence interval levels. Data were considered significant at P value <0.05.

2.3 RESULTS

2.3.1 PHYSICOCHEMICAL CHARACTERIZATION OF THE PRODUCED NPS

The prepared particles show absorption in the visible range owing to surface plasmon resonance at 405 nm and the TEM images show that the particles are spherical with homogenous size distribution (Figure 2.1a, b). As confirmed by the TEM images, the absorption spectra show one narrow band indicating the spherical shape of the obtained particles (Figure 2.2a, b).

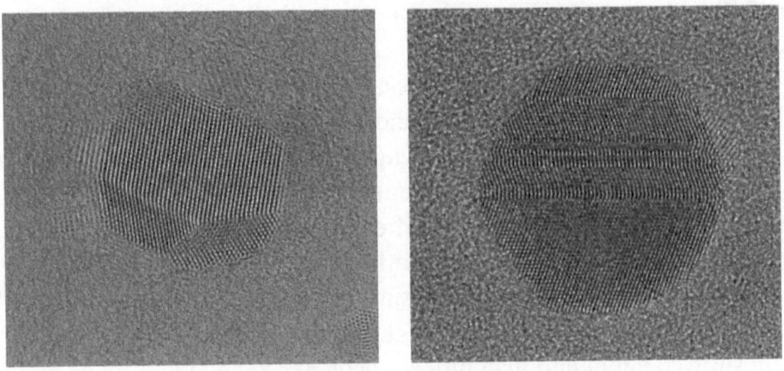

FIGURE 2.1 TEM micrographs of (a) AgNPs and (b) Co/AgNPs.

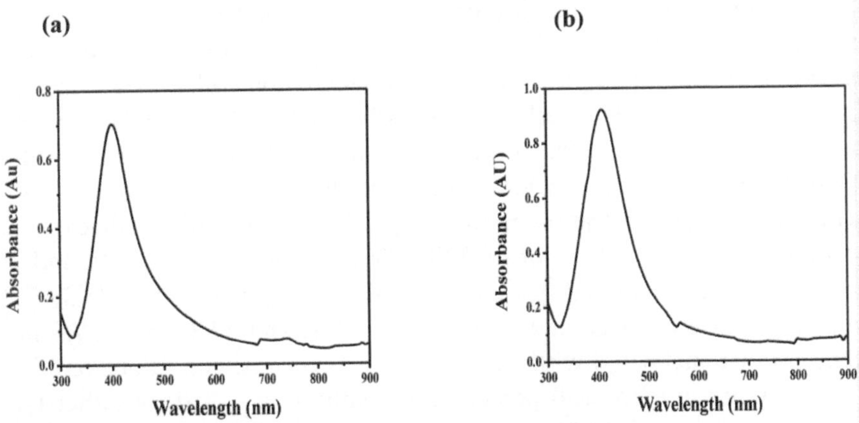

FIGURE 2.2 Absorption spectrum of (a) AgNPs (and (b) Co/AgNPs.

2.3.2 *IN VITRO* VIABILITY ASSAY

Our *in vitro* study was based on evaluation of the photo thermal cytotoxic effect of Ag and Co/Ag core shell nanoparticles upon treatment of HEp-2 laryngeal cancer cells.

A dark toxicity test was performed to investigate the effect of both AgNPs and Co/AgNPs on the viability of HEp-2 laryngeal cell carcinoma incubated with either type of nanoparticles in the absence of monochromatic light activation. Two different concentrations have been examined for their cytotoxic effect on HEp-2 tumor cells (10^{-6} M/L and 5×10^{-5} M/L) both at 5 and 24 h incubation times. Evaluation of cellular viability using NR assay resulted into absolutely no cell death at 10^{-6} M/L with either concentrations of both AgNPs or Co/AgNPs at 5 h incubation time. Longer incubation periods at 24 h, resulted into a slight decrease in cell viability as indicated by 3.13 percent and 5.29 percent, at 10^{-6} M/L of AgNPs and Co/AgNPs, respectively, and insignificant decrease in cell survival as shown by 13.56 percent and 15.84 percent at 5×10^{-5} M/L of AgNPs and Co/AgNPs, respectively. These results of dark toxicity test prove that both AgNPs and Co/AgNPs are not significantly toxic at either 10^{-6} M/L or 5×10^{-5} M/L (Figure 2.3a). Meanwhile, exposing HEp-2 laryngeal carcinoma cells to monochromatic blue light of 460 nm at different exposure times of 5, 10, and 15 min showed slight and insignificant decrease in cell viability as indicated by 0.88 percent, 4.83 percent and 7.8 percent, respectively (Figure 2.3b).

Treatment of HEp-2 cells with 5 x 10^{-5} M/L of either particles and 200 mW exposure to monochromatic blue light resulted into dramatic decrease in HEp-2 cells viability. The LC50 conditions of HEp-2 cells were exceeded and could not be detected upon incubation with either AgNPs or Co/AgNPs, as complete cell death was obtained at 10 min light exposure of AgNPs and Co/AgNPs, respectively (data not shown). On the other hand, HEp-2 laryngeal carcinoma cells that have been incubated for 24 h with 10^{-6} M/L concentration of either AgNPs or Co/AgNPs while exposed to monochromatic blue light of 200 mW and 460 nm for 5, 10, and 15 min resulted into gradual decrease in cell viability from 84.1, 75.3, 51.3 percent in case of cells incubated with AgNPs, and 77.3, 49.5, 37.8 percent in those incubated with Co/AgNPs. This means, conditions required for reaching the LC50 (50 percent cell death) at 10^{-6} M/L of either type of particles were 15 and 10 min light exposure for AgNPs and Co/AgNPs, respectively (Figure 2.4a, b).

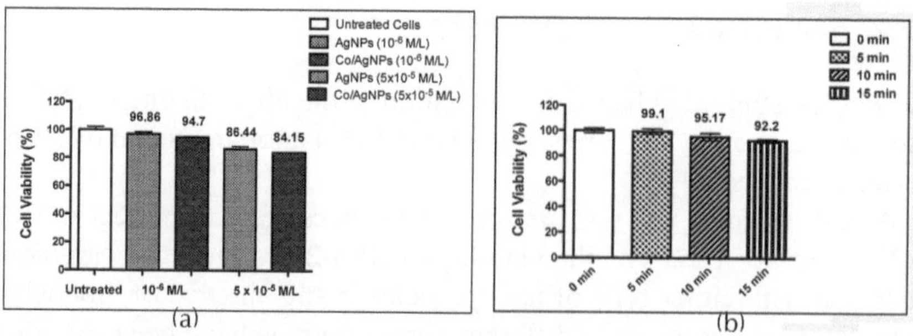

FIGURE 2.3(A) Dark toxicity test and (b) light toxicity test.

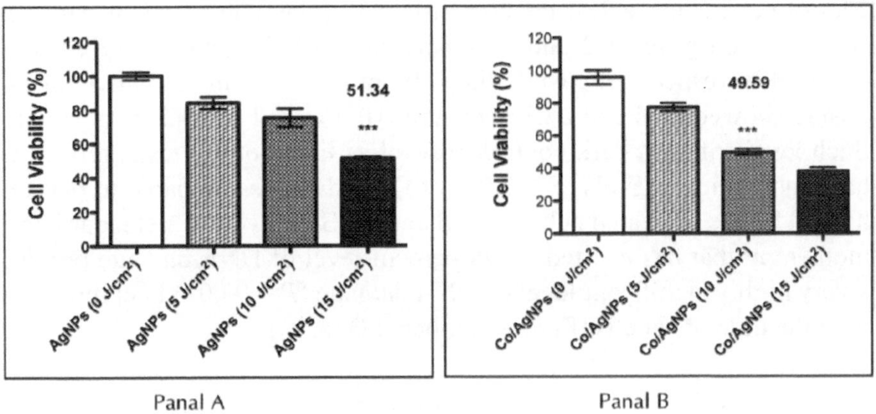

FIGURE 2.4 Cell viability test of HEp-2 laryngeal carcinoma cells incubated with 10^{-6} M/L (a) AgNPs and (b) Co/AgNPs and exposed to 200 mW monochromatic blue light of 460 nm.

2.3.3 EVALUATION OF GENOTOXICITY

In this chapter, an alkaline single-cell gel electrophoresis assay (Comet assay) was applied on peripheral blood lymphocytes to measure the single-strand breaks (SSB), double-strand breaks (DSB), as well as alkali labile sites as indicated by the number of DNA stand breaks for each therapeutic modality of either chemotherapy or PTT. The threshold level for cytotoxicity after application of each therapeutic modality was set at 70 percent cell viability, to minimize cytotoxicity-induced genotoxic effects [16]. The LC50 conditions of both AgNPs and Co/AgNPs concluded from the *in vitro* tumor viability assay in HEp-2 laryngeal carcinoma cells exerted less than 70 percent lymphocyte cell death as proved by Trypan Blue exclusion test. Accordingly, the genotoxicity test by alkaline comet assay was applied on those conditions (10^{-6} M/L of either type of NPs for 24 h incubation, and 15 min monochromatic blue light exposure) and compared with the untreated control cells (Figure 2.5, Panel A).

On one hand, blood lymphocytes treated with LC50 conditions of AgNPs gave significant percentage ($P < 0.05$) of DNA damage (49.85%) compared with the untreated control cells (Figure 2.5, Panel B (a)). On the other hand, lymphocytes treated with LC50 conditions of Co/AgNPs mediated PTT, resulted into a highly significant increase ($P = 0.01$) of DNA damage as indicated by 51.2 percent compared with the untreated control cells (Figure 2.5, Panel B (b)). Also, the sole effect of particles concentration gave 28.2 and 29.4 percent DNA damage for AgNPs and Co/AgNPs, respectively, while the 460 nm monochromatic blue light exposure showed 26.5 and 27.3 percent at 10 min and 15 min, respectively. Such levels of both dark control, as well as light control tests, proved to have insignificant levels ($P > 0.05$) of DNA damage compared to that exhibited by the untreated cells (23.4 percent). Bleomycin is a standard chemotherapy that has exerted 99.98 percent level of DNA damage proving a very highly significant level of DNA damage ($P = 0.001$) in comparison with the untreated cells (Figure 2.5, panel B (a, b)).

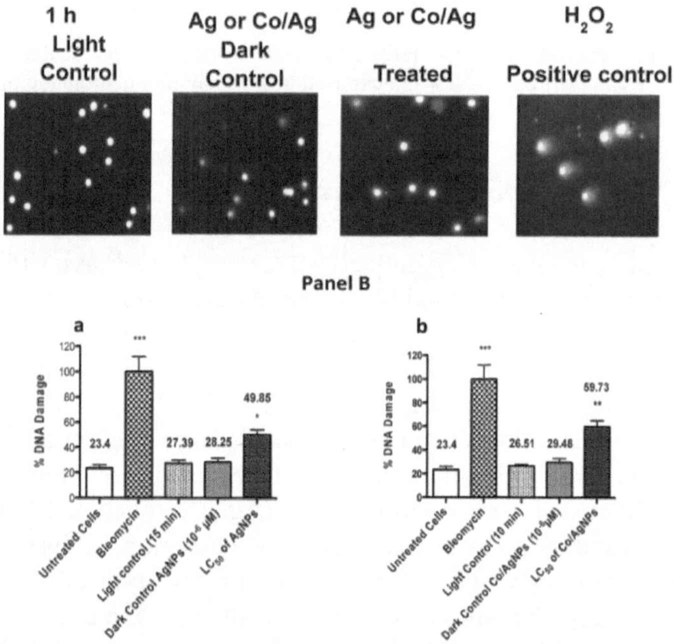

FIGURE 2.5 Genotoxicity Comet assay. Percentage of DNA damage exerted by LC50 conditions of AgNPs, panel (B (a)), and Co/AgNPs, panel (B (b)) on freshly isolated peripheral blood lymphocytes.

2.3.4 EVALUATION OF MUTAGENICITY

Chromosomal mutations resulted from treatment of peripheral blood lymphocytes with LC50 of either AgNPs or Co/AgNPs was investigated using standard karyotyping by GTG banding technique. Data of the light control and dark control, lymphocytes with LC50 conditions of either type of particles showed normal karyotype (46, XX) with no chromosomal aberrations (data not shown). On the other hand, metaphase analysis of lymphocytes treated with LC50 conditions of AgNPs and Co/AgNPs exhibited different types of mutations including chromatid type breaks as well as hypoploidy such as 45, XX, −18 and 45, XX, −18 in case of AgNPs treated lymphocytes, while 40, XX, −6, −10, −11, −12, −17, −20 chtb(1), and 45, XX, −22, chtb(9) in case of the Co/AgNPs-treated lymphocytes (Figure 2.6a, b).

Additionally, the mitotic index was investigated to evaluate the lethality magnitude of the LC50 of each drug on peripheral blood lymphocytes. Results proved significant decrease (53.3%) in the mitotic index of cells incubated with LC50 of AgNPs, and a highly significant decrease (40.37%) in mitotic index of cells treated with Co/AgNPs mediated PTT, in comparison with that of control cells, which have not been exposed to PTT. On the other hand, there has been a major drop in the mitotic index of blood lymphocytes treated for 24 h with the strong chemotherapeutic drug Bleomycin (4.33%) showing a very highly statistical significance ($P = 0.001$) compared to the control untreated cells (Figure 2.7).

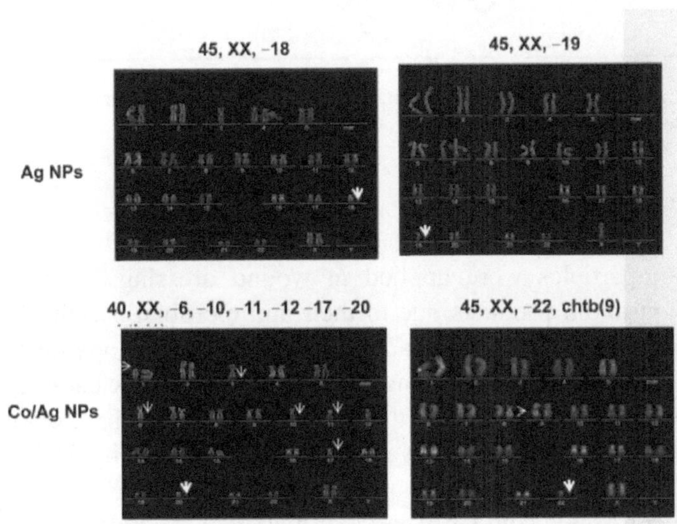

FIGURE 2.6 M-FISH analysis of human karyotype treated with LC50 conditions of either AgNPs or Co/AgNPs mediated PTT. Cells incubated with 10^{-6} M/L of either AgNPs, or Co/AgNPs mediated PTT, and exposed to monochromatic blue light of 460 nm and 200 mW.

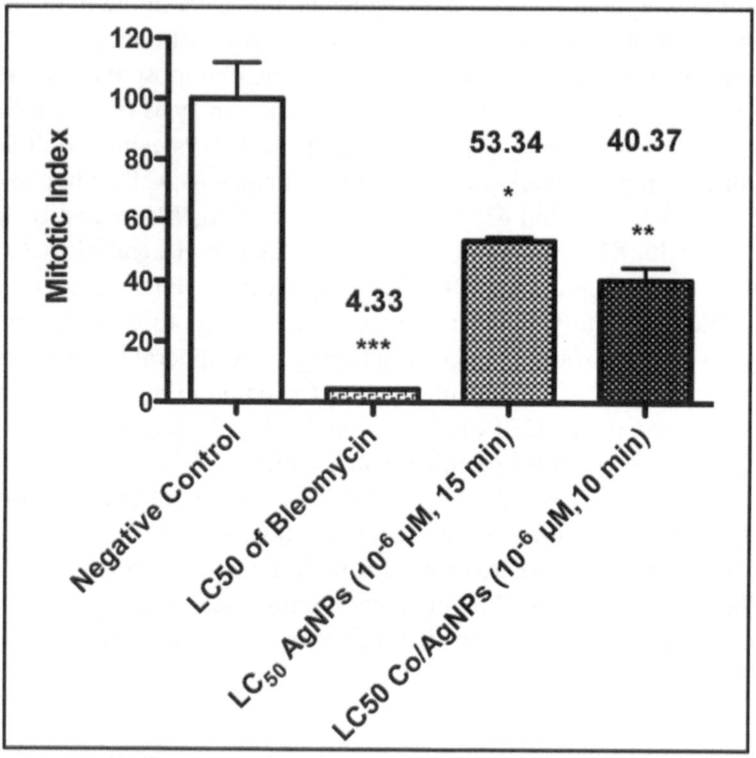

FIGURE 2.7 Effect of LC50 conditions of AgNPs and Co/AgNPs-mediated PTT on the mitotic index of peripheral blood lymphocytes.

2.4 DISCUSSION

Silver nanoparticles were applied in wound dressings, catheters, and various household products due to their antimicrobial activity [17]. The achievement of different sizes and shapes of silver nanoparticles can be obtained by minor modifications during their synthesis parameters [18, 19]. Therefore, in this chapter, the toxicity of AgNPs and Co/AgNPs on HEp-2 laryngeal cell line was evaluated using cell viability, DNA damage and chromosomal aberrations.

Spherical AgNPs and Co/AgNPs of average size 20 nm were used in this chapter. The optical absorption of both nanoparticles covers the spectral absorption range of 350–600 nm. The maximum emission wavelength of light emitting diode (LED) used in this chapter (460 nm) is found in the

AgNPs and Co/AgNPs absorption range. Both HEp-2 cells dark and light controls showed insignificant change when they were exposed to AgNPs and Co/AgNPs concentrations of 10^{-6} and 5×10^{-5} M/l or 5, 10, and 15 J/Cm^2 of light irradiance, respectively. This means that the reduction in the percentage of cell viability shown in Figure 2.4 compared with the untreated control cells is due to the activation of AgNPs and Co/AgNPs with the blue light photons (460 nm). Absorbed light increases the plasmon resonance of these nanoparticles, and by particles relaxation process the excess energy is dissipated in form of heat. The heat energy surrounding each nanoparticle causes cellular localized effect called hyperthermia, which is a state where cells absorb more heat than they can dissipate. Such a heat can be lethal to cancer cells [1].

Different light fluencies rates (5, 10, and 15 J/Cm^2) were applied in this chapter to have suitable LC50 of AgNPs and Co/AgNPs using the same above concentrations to be used in the cytotoxicity and genotoxicity studies in blood lymphocytes. Light influence rate of 15 J/Cm^2) in case of AgNPs treatment has the same reduction effect of the 10 J/Cm^2 in case of Co/AgNPs. This may reflect that Co/AgNPs is more sensitive than AgNPs to this kind of treatment. This sensitivity may be due to the fact that the thermal reaction deriving force in case of Co/AgNPs comes from the large reduction potential gap between the Ag^{+}/Ag and the Co^{+2}/Co redox couples which results in the consumption of Co cores and the formation of a hollow cavity of Ag nanoshells. The UV spectrum of this nanostructure exhibits a distinct difference from that of solid nanoparticles, which makes it a good candidate for application in photothermal materials [9].

Cellular uptake of nanoparticles via diffusion or endocytosis is followed by their random distributed everywhere in the cells. It has been reported that in case of normal human lung fibroblast cells (IMR-90) and human glioblastoma cells (U251), the transmission electron microscopic (TEM) analysis indicated the presence of AgNPs inside the mitochondria and nucleus, implicating their direct involvement in the mitochondrial toxicity and DNA damage [17]. This has encouraged us to study the side effect of the photothermal process of AgNPs and Co/AgNPs-treated blood lymphocytes.

In this chapter, we are interested in the genotoxic effects caused by the LC50 conditions (10^{-6} M/L NPs and 200 mW blue light) of either AgNPs or Co/AgNPs resulting from the viability assay. Genotoxicity was evaluated and showed percentage of DNA damage of a significant and highly significant level for the AgNPs and Co/AgNPs, respectively. This result is in agreement with the study made on Balb/3T3 cell model at which, statistically significant DNA damage of $AgNO^3$ (7–10 μM), and (>1 μM) Co nanoparticles was obtained indicating their strong genotoxic potential [20]. In another study, [21] the comet assay showed a statistically significant dose-related increase in percent Tail DNA for (10^{-5}–10^{-4} M) of CoNP ($P < 0.001$) upon 24 h incubation with peripheral blood lymphocytes. Meanwhile, a dose-dependent increase in genotoxicity has been observed when AgNPs were incubated with CHO-K1 cells causing 450 percent increase in DNA breakage at a concentration of 9×10^{-6} M compared with control cells ($P < 0.01$) [22]. In all cases, it is anticipated that DNA damage is augmented by deposition, followed by interactions of either silver or cobalt silver nanoparticles to the DNA leading to cell cycle arrest in the G_2/M phase [23].

Investigation of the effect of both AgNPs and Co/AgNPs on chromosomal aberrations of human peripheral blood lymphocytes indicated the exhibition of numerical (hypoploidy) chromosomal aberrations in case of AgNPs while both numerical (hypoploidy) and structural (Chromatid type breaks) aberrations in case of Co/AgNPs. Although no available studies have been found for the mutagenic effect of either AgNPs or Co/AgNPs on the level of human chromosomes, several studies have proved the mutagenic effect of both types of particles using the Ames mutagenicity test, as well as micronucleus test [22, 24].

Meanwhile, the genetic investigations showed significant drop (53.3%) in the mitotic index of cells incubated with LC50 of AgNPs and a highly significant decrease (40.37%) in mitotic index of cells treated with Co/AgNPs-mediated PTT in comparison with that of control cells, which have not been exposed to PTT. This suggests that PTT tumor cells killing could happen through a molecular mechanism involving DNA double strands breaks or chromosomal mutations. A possible mechanism could involve inhibition of extracellular signal-regulated kinases (ERKs) that are responsible for cell proliferation and differentiation, or down regulation of cyclin D1 and cyclin E which lead to antiproliferative effect of chlorophyll derivative [25, 26].

2.5 CONCLUSIONS AND RECOMMENDATIONS

In conclusion, AgNPs- and Co/AgNPs-mediated PTT can be used in the medical field offering a promising tool for cancer cells death. It provides considerable cytotoxic effect when administered by laryngeal cancer cells. However, both types of particles showed considerable genetic side effects both on DNA and chromosomal levels. Therefore, AgNPs- and Co/AgNPs-mediated PTT needs further optimization as well as *in vivo* investigations for possible future clinical applications.

The efficacy of AgNPs and Co/AgNPs should also be tested in other cancer cell lines, and their cellular uptake should be examined by confocal microscopy to specify the time required for their exocytosis that might influence their efficacy at tumor cell killing. Additionally, higher selectivity to tumor cells could be examined via *in vivo* applications, taking into consideration that the surface modifications of both types of nanoparticles might increase their specificity at tumor cell targeting.

ACKNOWLEDGMENT

The authors would like to thank Marwa Ramadan at the National Institute for Laser Enhanced Sciences—NILES for the synthesis of both AgNPs and Co/AgNPs. This work was been supported in part by the German Academic Exchange Service, *Deutsche Akademische Austausch Dienst (DAAD)*.

KEYWORDS

- **Ag and Co/Ag nanoparticles**
- **Genotoxicity**
- ***In vitro* study**
- **Mutagenicity**
- **Photothermal therapy**

REFERENCES

1. Van der Zee, J.; Heating the patient: A promising approach? *Ann. Oncol.* **2002**, *13(8)*, 1173–1184.
2. Kikumori, T.; Kobayashi, T.; Sawaki, M.; Anti-cancer effect of hyperthermia on breast cancer by magnetite nanoparticle-loaded anti-HER2 immunoliposomes. *Breast. Cancer. Res. Treat.* **2009**, *113*(3), 435–441.
3. Zheng, Y.; Hunting, D. J.; Ayotte, P.; Sanche, L.; Radiosensitization of DNA by gold nanoparticles irradiated with high-energy electrons. *Radiat. Res.* **2008**, *169(1)*, 19–27.
4. Zamiri, R.; Zakaria, A.; Husin, M. S.; Wahab, Z. A.; and Nazarpour, F. K.; Formation of silver microbelt structures by laser irradiation of silver nanoparticles in ethanol. *Int. J. Nanomed.* **2011**, *6*, 2221–2224.
5. Nam, Y. S.; Kang, H. S.; Park, J. Y.; Park, T. G.; Han, S. H.; and Chang, I. S.; New micelle-like polymer aggregates made from PEI-PLGA diblock copolymers: micellar characteristics and cellular uptake. *Biomaterials.* **2003**, *24(12)*, 2053–2059.
6. Iida, H.; Chemical Synthesis of Nanoparticles and their Applications to Bioanalysis and Medical Care. PhD Dissertation, Shinjuku, Tokyo, Japan: Waseda University; **2008**.
7. Ruddon R. W.; Cancer Biology. 4th edition, Oxford: Oxford University Press Inc.; **2007**.
8. Liau, S. Y.; Read, D. C.; Pugh, W. J.; Furr, J. R.; Russell, A. D.; Interaction of silver nitrate with readily identifiable groups: relationship to the antibacterial action of silver ions. *Lett. Appl. Microbiol.* **1997**, *25(4)*, 279–283.
9. Guildford, A. L.; Poletti, T.; Osbourne, L. H.; Di Cerbo, A.; Gatti, A. M.; and Santin, M.; Nanoparticles of a different source induce different patterns of activation in key biochemical and cellular components of the host response. *J. R. Soc. Interf.* **2009,** *6(41)*, 1213–1221.
10. Singh, N. P.; McCoy, M. T.; Tice, R. R.; and Schneider, E. L.; A simple technique 458 for quantitation of low levels of DNA damage in individual cells. *Exp. Cell Res.* **1988,** 184—91.
11. Aardema, M. J.; et al. Aneuploidy: A report of an ECETOC task force. *Mutat. Res.* **1998**, *410(1)*, 3–79.
12. Anderson, D.; and Plewa, M. J.; The International Comet Assay Workshop. *Mutagenesis.* **1998**, *13(1)*, 67–73.
13. Preston, R. J.; San Sebastian, J. R.; and McFee, A. F.; The *in vitro* human lymphocyte assay for assessing the clastogenicity of chemical agents. *Mutat. Res.* **1987**, *189(2)*, 175–183.
14. Liehr, T.; et al. Direct preparation of uncultured EDTA-treated or heparinized blood for interphase FISH analysis. *Appl. Cytogenet.* **1995** *21(6)*, 185–188.
15. Chudoba, I.; Plesch, A.; Lörch, T.; Lemke, J.; Claussen, U.; and Senger, G.; High resolution multicolor-banding: A new technique for refined FISH analysis of human chromosomes. *Cytogenet. Cell. Genet.* **1999**, *84(3–4)*, 156–160.
16. Hartmann, A.; Plappert, U.; Poetter, F.; and Suter, W.; Comparative study with the alkaline Comet assay and the chromosome aberration test. *Mut. Res.* **2003**, *536(1–2)*, 27–38.

17. Pal, S.; Tak, Y. K.; and Song, J. M.; Does the antibacterial activity of silver nanoparticles depend on the shape of the nanoparticle? A study of the gram-negative bacterium escherichia coli. *Appl. Environ. Microbiol.* **2007**, *73(6),* 1712–1720.
18. Bogle, K. A.; Dhole, S. D.; and Bhoraskar, V. N.; Silver nanoparticles: Synthesis and size control by electron irradiation. *Nanotechnology*. **2006**, *17(13),* 3204.
19. Panáček, A.; Kvitek, L.; Prucek, R.; Kolar, M.; Vecerova, R.; Pizurova, N.; and Zboril, R.; Silver colloid nanoparticles: Synthesis, characterization, and their antibacterial activity. *J. Phys. Chem. B.* **2006**, *110(33),* 16248–16253.
20. Munaro, B.; Mechanistic in vitro tests for genotoxicity and carcinogenicity of heavy metals and their nanoparticles. PhD Dissertation, Germany: Konstanz University; **2009**.
21. Colognato, Bonelli, A.; Ponti, J.; Farina, M.; Bergamaschi, E.; and Sabbioni, E.; Comparative genotoxicity of cobalt nanoparticles and ions on human peripheral leukocytes *in vitro. Mutagenesis.* **2008**, *23(5),* 377–382.
22. Kim, H. R.; Park, Y. J.; Shin, D. Y.; Oh, S. M.; and Chung K. H.; Appropriate. *In vitro* methods for genotoxicity testing of silver nanoparticles. *Environ. Health Toxicol.* **2013**, *28,* 1–8.
23. AshaRani, P. V.; Mun G. L. K.; Hande M. P.; and Valiyaveettil, S.; Cytotoxicity and genotoxicity of silver nanoparticles in human cells. *ACS Nano*. **2009,** *3(2),* 279–290.
24. Yan, L. I.; et al. Genotoxicity of silver nanoparticles evaluated using the ames test and in vitro micronucleus assay. *Mut. Res.* **2012**, *745(1–2),* 4–10.
25. Chiu, L. C.; Kong, C. K.; and Ooi, V. E.; Antiproliferative effect of chlorophyllin derived from a traditional Chinese medicine Bombyxmori excreta on human breast cancer MCF-7 cells. *Int. J. Oncol.* **2003**, *23(3),* 729–735.
26. Chiu, L. C.; Kong, C. K.; and Ooi, V. E.; The chlorophyllin-induced cell cycle arrest and apoptosis in human breast cancer MCF-7 cells is associated with ERK deactivation and cyclin D1 depletion. *Int. J. Mol. Med.* **2005**, *16(4),* 735–740.

CHAPTER 3

A LECTURE NOTE ON APPLICATION STABLE RADICALS FOR STUDY OF BEHAVIOR OF BIOLOGICAL SYSTEMS

M. D. GOLDFEIN and E. G. ROZANTSEV

CONTENTS

3.1 INTRODUCTION

The presence of paramagnetic particles in liquid or solid objects opens new opportunities of their studying by the Electron paramagnetic resonance (EPR) technique. Ready free radicals and substances forming paramagnetic solutions owing to spontaneous homolization of their molecules in liquid and solid media (such as triphenylmethyl dimer, Frémy's salt, or 4,8-diazaadamantan-4,8-dioxide) can act as sources of paramagnetic particles.

The experimental technique of radio spectroscopic examination of condensed phases with the aid of paramagnetic impurities is usually called the paramagnetic probe method. Though iminoxyl radicals have found broadest applications for probing of biomolecules; nevertheless, the first application of the paramagnetic probe technique to study a biological system is associated with a quite unstable aminazine radical cation:

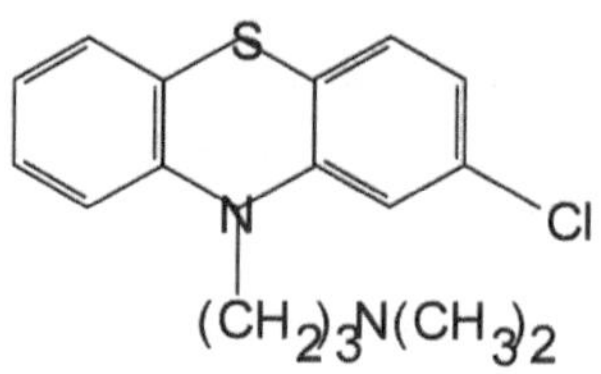

The progress in the theory and practice of EPR usage in biological research is restrained by the narrow framework of chemical reactivity of nonfunctionalized stable radicals with a localized paramagnetic center like

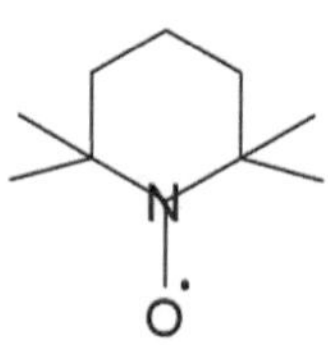

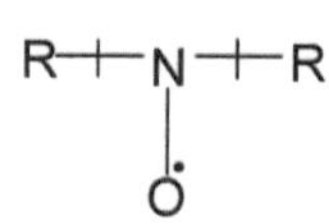

The substances of this class only enter into common, well-known free radical reactions, such as recombination, disproportionation, addition to multiple bonds, isomerization, and β-splitting [1]. All these reactions proceed with the indispensable participation of a radical center and steadily lead to full paramagnetism loss. And still the synthesis of nonfunctionalized stable radicals plays a very important role. No expressed delocalization of an uncoupled electron over a multiple-bond system has been shown to be obligatory for a paramagnetic to be stable.

Despite of the basic importance of the discovery of stable radicals of a nonaromatic type [2], this event has not changed contemporary ideas on the reactivity of stable radicals.

In the early 1960s, one of the authors of this book laid the foundation of a new lead in the chemistry of free radicals, namely, the synthesis and reactivity of functionalized stable radicals with an expressed localized paramagnetic center [3].

The opportunity to obtain and study a wide range of such compounds with various functional substituents arose in connection with the discovery of free radical reactions with their paramagnetic center unaffected.

Functionalized free radicals have found broad applications as paramagnetic probes for exploring molecular motion in condensed phases of various natures. The introduction of a spin label technique (covalently bound paramagnetic probe) is associated with their usage; its idea is not new and is based on the dependence of the EPR spectrum shape of the free radical on the properties of its immediate atomic environment and the way of interaction of the paramagnetic fragment with the medium. The reactions of free radicals with their paramagnetic center unaffected (Neumann–Rozantsev's reactions) [4] have became the chemical basis of obtaining spin-labeled compounds.

The concept of the usage of nonradical reactions of radicals to study macromolecules was formulated at the Institute of Chemical Physics (USSR Academy of Sciences) by Lichtenstein [5], and the theoretical bases of this method, calculation algorithms for the correlation times of rotary mobility of a paramagnetic particle from their EPR spectrum shape were developed by McConnell [6], Freed and Fraenkel [7], Kivelson [8], and Stryukov [9]. Have investigated the behavior of iminoxyl radicals in various systems and obtained important and interesting results [10].

Let us cite several important aspects of the application of organic paramagnets to researching biological systems.

3.2 APPLICATION OF IMINOXYL FREE RADICALS FOR STUDYING OF IMMUNE GAMMA GLOBULINS

From the physicochemical viewpoint, the mechanism of various immunological reactions is determined by changes in the phase state of a system.

Despite of the wide use of these reactions in medical practice, the nature of interaction of antigens with antibodies is not quite clear. To study this process, gamma globulins labeled with iminoxyl radicals were used [11]. The molecule of gamma globulin is known to consist of four polypeptide chains bound with each other with disulfide bridges. When the interchain disulfide bonds are split, the polypeptide chains continue to keep together.

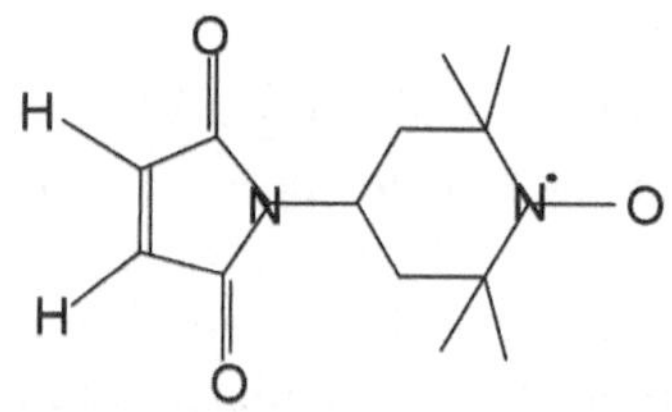

A spin label (an iminoxyl derivative of maleimide) was attached to the sulfhydryl groups obtained by restoration of disulfide bonds with β-mercaptoethanol. Experiments were made on the rabbit and human gamma globulins. The EPR spectra in both cases corresponded to rather high mobility of free radicals (correlation times $\tau = 1.1 \cdot \times 10^{-9}$ s for human gamma globulin and $7.43 \cdot \times 10^{-9}$ s for rabbit's one). By comparing the correlation times in these proteins with the values obtained in experiments with serum albumin labeled with sulfhydryl groups, treated with urea ($\tau = 1.09 \cdot \times 10^{-9}$ s) and dioxane ($2.04 \cdot \times 10^{-9}$ s), it is possible to conclude that the fragments of polypeptide chains bearing free radicals possess no ordered secondary structure. This is in accord with data on very low contents of α-helical structures in gamma globulins. Such a character of the EPR spectrum of immune gamma globulin with preservation of its specific activity opens the possibility to explore conformational and phase transitions at specific antigen–antibody reactions. Sharp distinctions in the mobility of spin labels were revealed at sedimentation of the rabbit antibodies by salting-out with ammonium sulfate and precipitation with a specific antigen (egg albumin). Precipitation of antibodies with a specific antigen led only to a small reduction of the paramagnetic label mobility whereas sedimentation with ammonium sulfate caused strong retardation of the rotary mobility of free radicals (Figure 3.1).

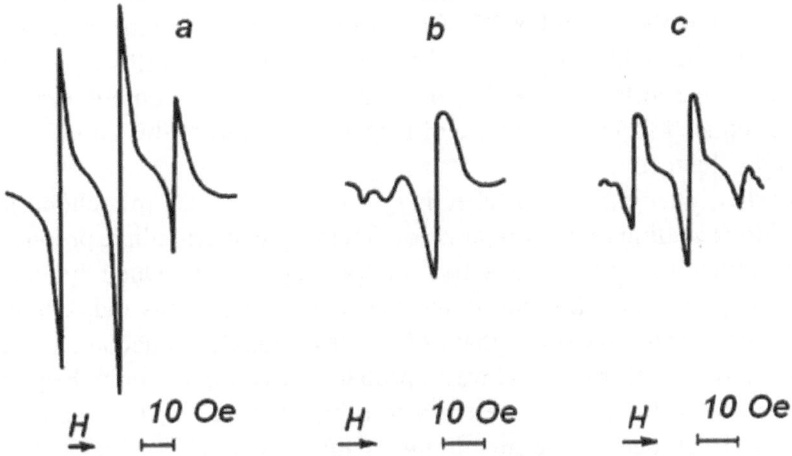

FIGURE 3.1 EPR spectra of gamma globulin labeled with an iminoxyl radical with its free valency unaffected: (a) in solution; (b) in the precipitate obtained by salting-out with ammonium sulfate; (c) in the specific precipitate.

These results can be considered as direct confirmation of the alternative theory [12] according to which the precipitate formation is associated with the immunological polyvalency of the antigen and antibody relative to each other. Really, excepting the location of spin labels inside the antibody's active center, it is possible to conclude that the rather mobile condition of spin labels in the antigen–antibody precipitate may remain, only if there is no strong dehydration of the antibodies owing to intermolecular interactions. Unlike gamma globulin precipitated with ammonium sulfate, the specific precipitate, according to the lattice theory, has a microcellular structure. At long storage of the antigen–antibody precipitate with no addition of stabilizers, the mobility degree of iminoxyl radicals sharply decreased. This is apparently a result of secondary dehydration of the antibodies owing to protein molecule interactions in the precipitate.

3.3 EXPLORING STRUCTURAL TRANSITIONS IN BIOLOGICAL MEMBRANES

Biological membranes, in particular, mitochondrial membranes, are known to play a huge role in redox processes in the cell, being the very place of respiratory chain enzyme localization. Baum and Riske [13] have found essential distinctions in the properties of one of the mitochondrial membrane fragments (Complex III of Electron Transport Chain) upon transition from the oxidized form to the reduced one: the sulfhydryl groups, easily titrated in the oxidized form of this fragment, become inaccessible in the reduced form. The nature of trypsin digestion of this complex also strongly changes.

Earlier, changes in the repeating structural units of mitochondria in conditions leading to the formation of macroergic intermediate products or their provision, for example active ion transfer, were revealed by electron microscopy [14]. All this has allowed us to assume that any redox reaction catalyzed by an enzymatic chain of electron transfer is accompanied by some kind of conformational wave, probably covering not only the protein component of the membrane but also a higher level of its organization, namely: fragments of the membrane of more or less complexity degree, including its lipidic part. To verify this assumption, a modification of the spin label method (a method of noncovalent bound paramagnetic probe) was used. The radical is kept by the matrix (membrane) involving only weak hydrophobic bonds. Such an approach allows studying of weak interactions in the system without essential disturbance of the biochemical functions of the biomembrane and its structure. The paramagnetic probe was 2,2,6,6-tetramethylpiperidine-1-oxyl caprylic ester:

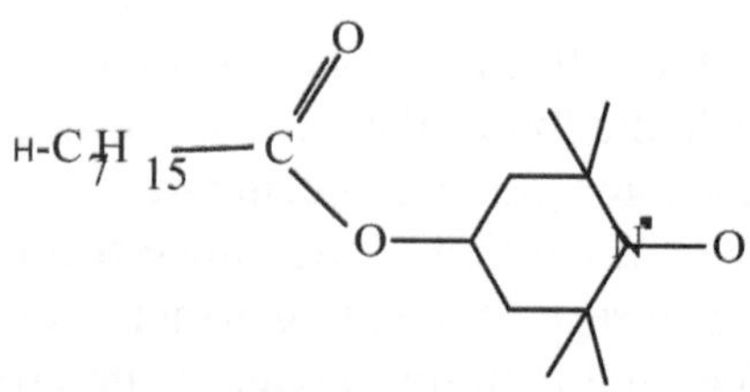

This compound was prepared from caprylic acid chloranhydride and 2,2,6,6-tetramethyl-4-oxpiperidin-1-oxyl in a triethylamine medium by a radical reaction with free valency unaffected. The paramagnetic probe was introduced into a suspension of electron transport particles (ETPs) isolated from the bull heart mitochondria by the technique described in [14]. at the Laboratory of Bioorganic Chemistry, Moscow State University. These fragments of the mitochondrial membrane are characterized by a rather full set of enzymes of the respiratory chain with the same molar ratio as in the intact mitochondria [14]. The ability of oxidizing phosphorylation, however, is lost under the used way of isolation.

The paramagnetic probe is insoluble in water but solubilized by ETP suspended in a buffer solution. Therefore, the observed EPR spectrum is free of any background as the radicals are not attached with the object under study. The presence of a voluminous hydrocarbonic chain provides embedding of a molecule of the probe into the lipidic part of ETP. Therefore, the EPR spectrum reflects the condition of exactly this fraction of the membrane. To detect conformational transitions, EPR spectra were recorded before and after the introduction of oxidation substrata (succinate and NAD-N), and after oxidation of the earlier reduced respiratory chain with potassium ferricyanide. Typical results are shown in Figure 3.2 [14].

The enhanced anisotropy of the EPR spectrum of iminoxyl after substratum introduction is clearly seen. The spectrum in the ferricyanide-oxidized ETP almost does not differ from that in the intact ETP.

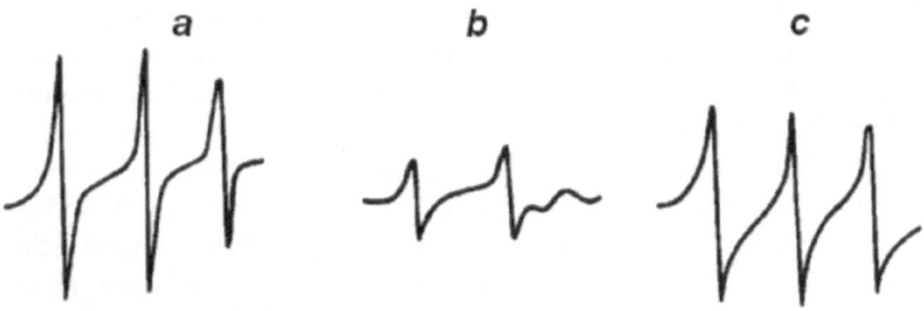

FIGURE 3.2 Changes in the EPR spectrum anisotropy of the hydrophobic iminoxyl radical in a suspension of electron transport particles from bovine heart mitochondria with oxidation substrates added.

ETP inactivation by long storage at room temperature or cyanide inhibition eliminated this effect. Comparison of the shape of signals and correlation times shows that the EPR spectrum of iminoxyl in the intact ETP consists of two signals differing by anisotropy. The radical localized in that part of the membrane where the effective free volume available for radical motion is rather large gives a weakly anisotropic signal. The strongly anisotropic (retarded) spectrum belongs to the radicals localized in other sites of the system with a smaller effective free volume. Reduction of the respiratory chain with substrata leads, owing to cooperative-type conformational transitions, to a reduced fraction of sites with large free volume (i.e., to an increased microviscosity of the immediate environment of the radical). The correlation time of the whole spectrum changes from 20×10^{-10} s in the oxidized ETP to 4×10^{-10} s in the reduced ETP.

Concurrently with enhancing the anisotropy of the signal, its intensity decreases as well as iminoxyl reduces, apparently, to hydroxylamine derivatives. Potassium ferricyanide inverts this process. It is necessary to consider that oxidation substrata, themselves, do not interact considerably with iminoxyls. Obviously, the conformational transition not only leads to a changed microviscosity but eliminates any steric obstacles complicating reduction of the radical. In principle, this circumstance points to possibly a new, actually chemical, aspect of application of the paramagnetic probe technique.

3.4 EXPLORING THE STRUCTURE OF SOME MODEL SYSTEMS

The application of the paramagnetic probe technique in systems like biological membranes poses a number of questions concerning the behavior of hydrophobic labels in media with an ordered arrangement of hydrophobic chains. As the first stage, mixtures of a nonionic detergent (Tween 80) and water were studied. Tween 80 originates from polyethoxylated sorbitan and oleic acid (polyoxyethylene (20) sorbitan monooleate) and classifies as a nonionic detergent based on polyethylene oxide [14]. Our choice of this object was determined by some methodical conveniences, and also some literature data on the structure of aqueous solutions of Tween, obtained by classical methods (viscometry, refractometry, etc.) The properties of this detergent are also interesting in themselves since it finds quite broad applications for biological membrane fragmentation.

The esters of 2,2,6,6-tetramethyl-4-oxypiperidin-1-oxyl and saturated acids of the normal structure with a hydrocarbonic chain length of 4, 7, or 17 carbon atoms or the corresponding amides were used as a paramagnetic probe. For comparison, the behavior of hydrophobic labels IV and V in these systems was also studied.

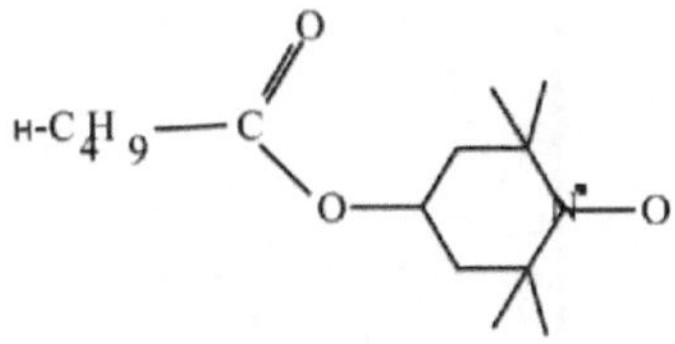

I

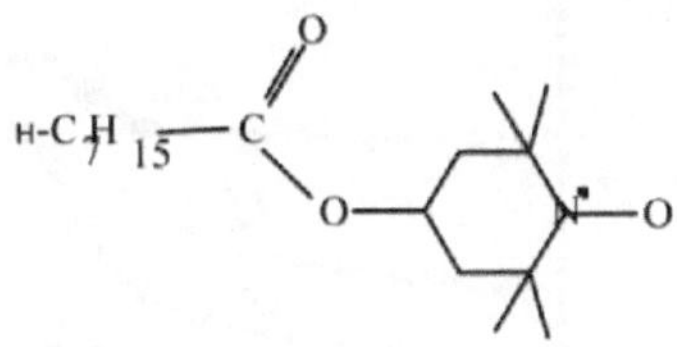

II

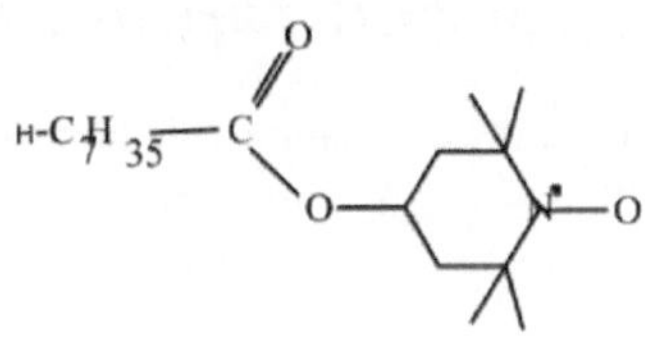

III

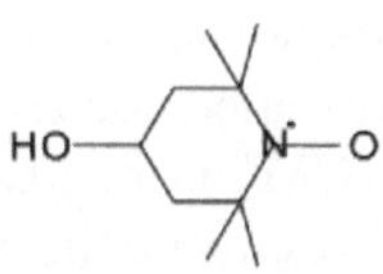

IV

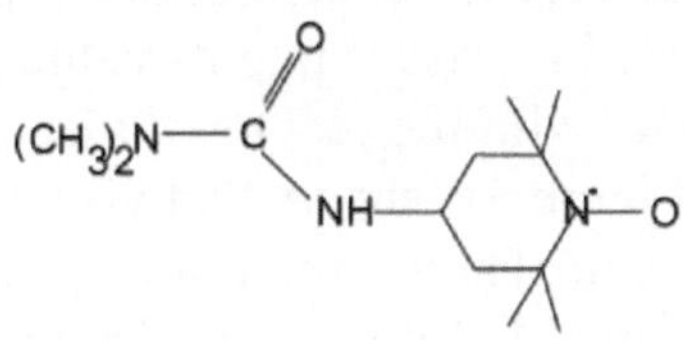

V

The course of changes in the correlation time of radical rotation as a function of the Tween 80 concentration is shown in Figure 3.3.

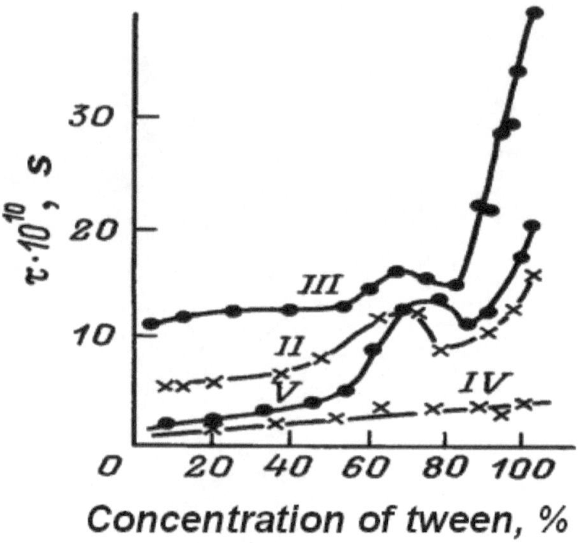

FIGURE 3.3 Changes of the correlation time in the detergent–water system, for nitroxyl radicals with a strongly localized paramagnetic center.

It is possible to resolve several areas, apparently, corresponding to various structure types. The initial fragment, only distinguishable for the easiest radicals, corresponds to an unsaturated Tween solution in water. This region is better revealed for the detergents with a higher critical micelle concentration (CMC), for example, for sodium dodecyl sulfate (Figure 3.4) [15, 16]. Then micelle formation occurs. The correlation time of water-insoluble labels increases and comes to a plateau; at further increasing detergent concentration, it passes through a maximum, then monotonously increases up to its value in pure Tween. It is useful to compare these data with the results of viscosity measurements by usual macromethods. Figure 3.5 shows that viscosity has one extremum about 60 percent of Tween. Therefore, at a high Tween concentration, the effective volume available for probe molecule rotation and macroviscosity do not correlate.

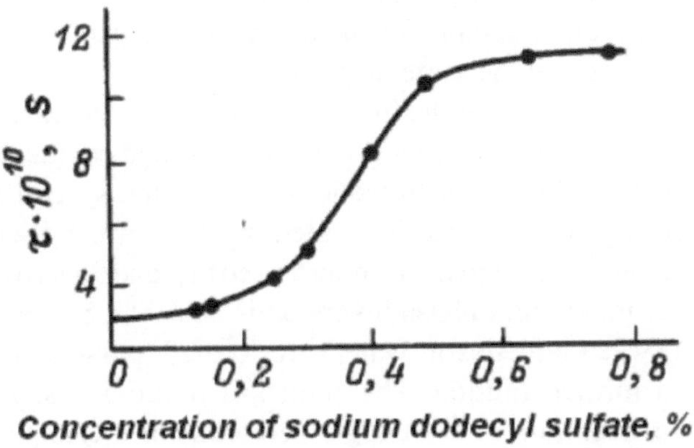

FIGURE 3.4 Estimation of the critical micelle concentration of sodium dodecylsulfate with the aid of 2,2,6,6-tertamethyl-4-hydroxypiperidyl-1-oxyl varerate.

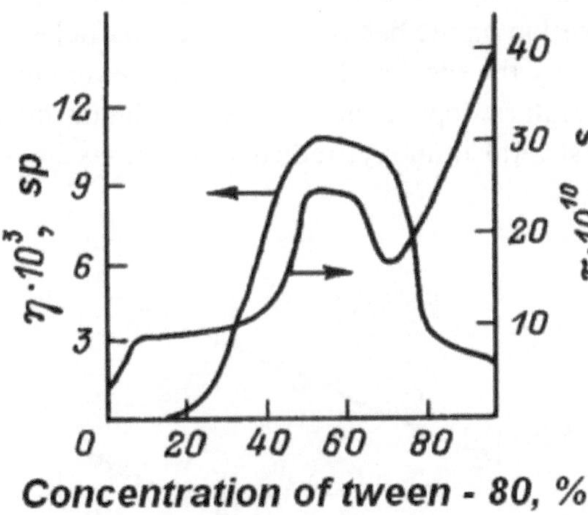

FIGURE 3.5 Changes in the microviscosity and macroviscosity of the water–Tween 80 system.

These results suggest the following interpretation. In pure Tween, the lamellar structure provides easy layer-by-layer sliding, hence, the macroviscosity of the system is low. However, in the absence of water, the interaction of the polar groups is strong, the hydrocarbonic chains are ordered, and the effective free volume in the field of radical localization is small. Small water amounts lead to the formation of defects in the layered structure. Sliding is hindered, and the viscosity increases. However, moistening breaks the close interaction of the polar groups in Tween. These groups are deformed, at the same time, the hydrocarbonic chains are disordered. The microviscosity of the hydrocarbonic layer decreases. Upon termination of hydration of the polar groups, water-filled cavities are formed. They are a structural element (micelle) of which the system is built, for example, a hexagonal P-lattice is formed, by Luzzatti. Structure formation manifests itself as increasing microbiscosity and macroviscosity. In the field of the maximum, phase inversion is possible. Structural units of a new type (Tween micelles in water) are formed, passing into colloidal solution upon further dilution. The course of microviscosity changes at high Tween concentrations amazingly resembles the change in correlation time when some lyophilized cellular organelles are moistened. This similarity confirms that in the field of τ maximum, where restoration of the biochemical activity of chloroplasts begins, a phase transition occurs of the same type as in the LC detergent–water systems.

Some information on the behavior of radical particles in colloidal systems is provided by the results of temperature measurements. Figure 3.6 shows the temperature dependence of the correlation time in Arrhenius' coordinates for several iminoxyl radicals of various hydrophobicity degrees.

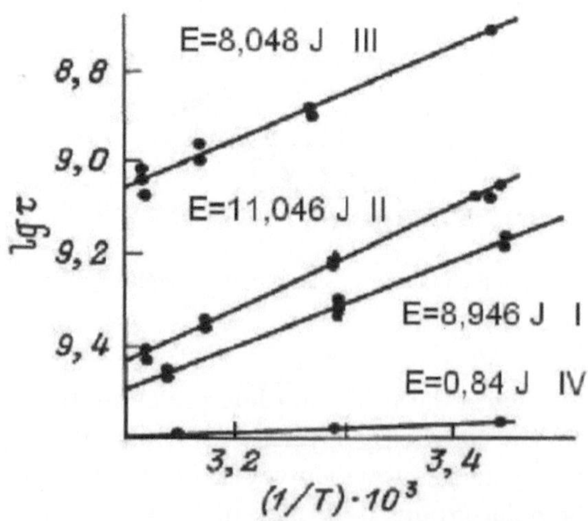

FIGURE 3.6 Activatyion energy of rotational diffusion of free iminoxyl (nitroxyl) radicals in the Tween–water (30% of Tween 80) system.

The strong dependence of the preexponential factor on chain length, apparently, confirms that the observed correlation time really reflects rotation of the radical molecule as a whole. However, this result also allows another interpretation, namely: depending on the hydrocarbonic chain length, the radical introduces into a detergent micelle more or less deeply. This may lead to a changed rotation frequency of iminoxyl groups round the ordinary bonds in the molecule, depending on the environment of the polar end of the radical. If the polar group of the radical is on the micelle-solvent interface, the measured frequency should depend on the surface charge (potential) of the micelle. In our opinion, this circle of colloid chemical problems, closely connected with questions of transmembrane transfer in biological systems, will provide one more application field of the paramagnetic probe technique.

3.5 SOLVING OTHER BIOLOGICAL PROBLEMS WITH STABLE RADICALS

Further progress in the field of the usage of stable paramagnets to solve various biological problems is reflected in numerous reviews and monographs [17–20].

For dynamic biochemistry, of undoubted interest are local conformational changes of protein molecules in solutions. The distances between certain loci of biomacromolecules, in principle, can be estimated quantitatively by means of stable paramagnets. Upon introduction of iminoxyl fragments into certain sites of native protein (NRR-method) the distance between the neighboring paramagnetic centers can be calculated from the efficiency of their dipole–dipole interaction in vitrified solutions of a spin-labeled preparation (the EPR method).

The first attempt to estimate the distances between paramagnetic centers in spin-labeled mesozyme and hemoglobin was undertaken by Liechtenstein [21], who indicated prospects of such an approach. In this regard, there appeared a need of identification of a simple empirical parameter in EPR spectra for quantitative assessment of the dipole–dipole interaction of paramagnetic centers.

A convenient empirical parameter was found when studying vitrified solutions of iminoxyl radicals. It was the ratio of the total intensity of the extreme components of a spectrum to the intensity of the central one

(Figure 3.7). To establish a correlation of the d_1/d value with the average distance between the localized paramagnetic centers, the corresponding calibration plots are drawn (Figure 3.8). Calculations have shown that the d_1/d parameter depends on the value of dipole–dipole broadening, being in fair agreement with independently obtained experimental results.

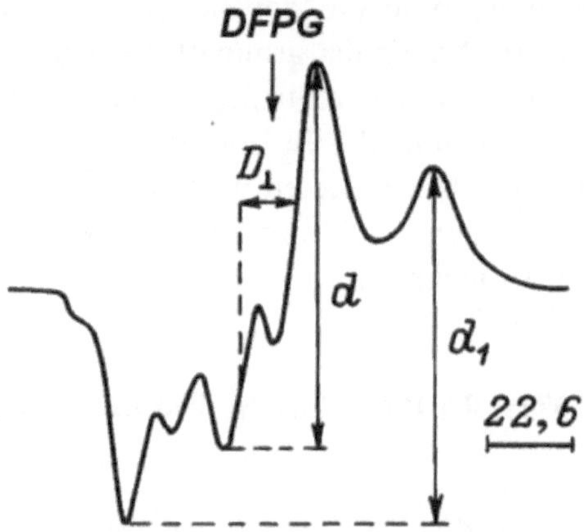

FIGURE 3.7 EPR spectrum of the iminoxyl free biradical (2,2,6,6-tetramethyl-4-hydroxypiperidine-1-oxyl phthalate) vitrified in toluene at 77 K.

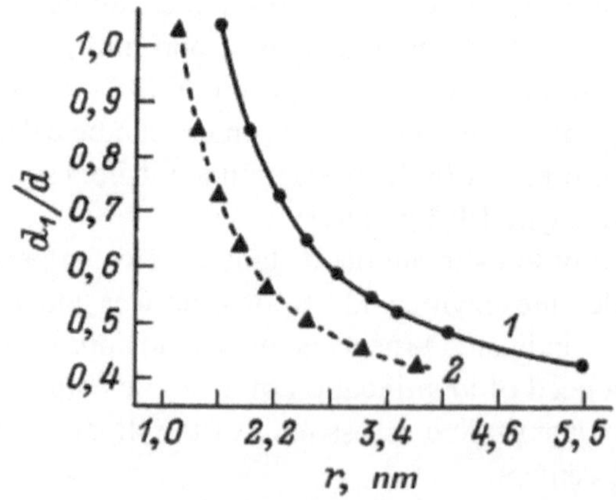

FIGURE 3.8 Dependence of the d_1/d paremeter of an EPR spectrum on the mean distance r between interaction paramagnetic centers of iminoxyl radicals (1) and biradicals (2) at 77 K.

Subsequently, the methods of quantitative assessment of the distances between paramagnetic centers in biradicals and spin-labeled biomolecules became reliable tools for structural research.

When determining relaxation rate constants of various paramagnetic centers in solution, the values of these constants have appeared to significantly depend on the chemical nature of the functional groups in iminoxyl radicals.

Sign inversion of the electrostatic charge of the substituent and its distance from the paramagnetic iminoxyl group have the strongest impact on the values of the constants. The electrostatic effect caused by the value and sign of the charge of paramagnetic particles interacting in solution is a more essential factor. An increased ionic strength leads to a change in the k value in qualitative agreement with Debye's theory. The substituent's mass is another factor considerably influencing the value of the constant. If the substituent is a protein macromolecule, the value of k decreases twice. The possibilities of application of the paramagnetic probe technique for detection of anion and cation groups and estimation of distances to explore the microstructure of protein were analyzed. The dependences obtained in experiment show that Debye's equation with $D = 80$ can be successfully applied to experimental data analysis, and, in particular, to estimating the distance from the iminoxyl group of a spin label on protein to the nearest charged group if this distance does not exceed 1.0–1.2 nm.

Considering the role of free radical processes in radiation cancer therapy, it was offered to investigate the influence of iminoxyl radicals upon the organism of laboratory animals. Pharmacokinetic studies [22] have shown that the elementary functionalized derivatives of 2,2,6,6-tetramethylpiperidin-1-oxyl possess rather low toxicity and show an expressed antileukemic activity. The highest values of retardation coefficients were observed for hematopoietic stem cells in peripheral blood and marrow. This circumstance stimulated further structural and synthetic research directed to obtaining more effective and less toxic cancerolytics and sensibilizers for radiation cancer therapy.

After the first publication, for a rather short time, many potential cancerolytics were synthesized in the Laboratory of Stable Radicals, Institute of Chemical Physics, USSR Academy of Sciences, among which of greatest interest from biologists were so-called paramagnetic analogs of some known antitumor preparations (e.g., a paramagnetic analog of Thiotepa) [22]:

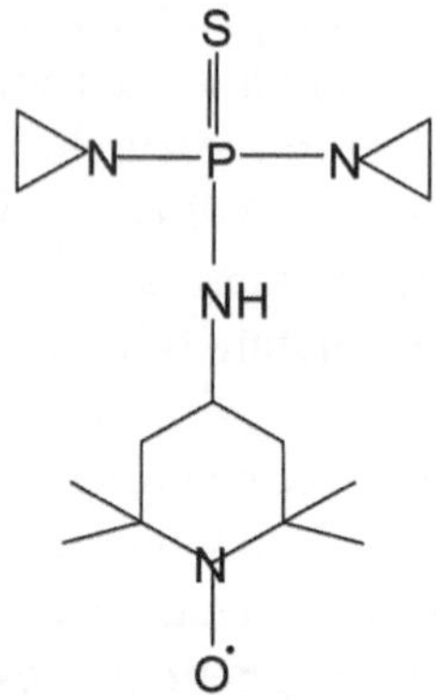

A brief review on the usage of stable paramagnets of the iminoxyl series in tumor chemotherapy is presented by Suskina [23].

KEYWORDS

- **Biology**
- **Organic paramagnetic method**
- **Radio spectroscopy**
- **Sensing, mechanism**
- **Study**

REFERENCES

1. Rozantsev E. G.; Chemical Encyclopedic Dictionary. M. SE. **1983**, 489 p. [in Russian]
2. Rozantsev E. G.; Lebedev O. L.; and Kazarnovskii S. N.; Diploma for the Opening Number 248 of 05.10.1983. TS. **1982**, *6,* 6 p. [in Russian]
3. Rozantsev E. G.; Doctor Dissertation. Moscow: Institute of Chemical Physics, USSR Academy of Sciences; **1965**, [in Russian].
4. Zhdanov R. I.; Nitroxyl radicals and non-radical reactions of free radicals. Bioactive Spin Labels. Zhdanov, R. I.; ed. Berlin, Heidelberg, New York: Springer; **1991**, 24 p.
5. Liechtenstein, G. I.; Method of spin labels in molecular biology. M.: Nauka, **1974**, 255 p. [in Russian]
6. McConnell, N. J.; *Chem. Phys*. **1956**, *25,* 709.
7. Freed, J.; and Fraenkel, G. J.; *Chem. Phys*. **1963**. *39,* 326.
8. Kivelson D.; *J. Chem. Phys*. **1960,** *33*, 4094.
9. Stryukov V. B.; Stable radicals in chemical physics. M.: Znanie, **1971**, [in Russian].

10. Goldfeld, M. G.; Grigoryan, G. L.; and Rozantsev, E. G.; Polymer (Gath. of preprints. M. ICP AS SSSR. **1979**, 269 p. [in Russian]
11. Grigoryan, G. L.; Tatarinov S. G.; Cullberg L. Y.; Kalmanson A. E.; Rozantsev E. G.; and Suskina V. I.; Abstract of USSR Academy of Sciences. **1968**, *178, 230(31),* 768 p. [in Russian]
12. Pressman, D.; Molecular Structure and Biological Specificity. Washington. D. C., **1957**.
13. Baum, H.; Riske, J.; Silman, H.; and Lipton, S.; *Froc. Nat. Acad. Sci. US.* **1967**, *57,* 798.
14. Rozantsev, E. G.; Biochemistry of meat and meat products (General Part). Ed. Manual for students. M. Depi. print. **2006**, 240 p. [in Russian]
15. Goldfeld, M. G.; Koltover, V. K.; Rozantsev, E. G.; Suskina V. I.; Kolloid. Zs. **1970**.
16. Schenfeld, H.; Nonionic Detergents. **1963**.
17. Smith, J.; Shrier–Muchillo, Sh.; and Murch, D.; Method of spin labels. Free radicals in biology. M.: Mir. **1979,** *1*, 179.
18. Method of Spin Labels and Probes, Problems and Prospects. M.: Nauka, **1986**, [in Russian].
19. Nitroxyl Radicals. Synthesis, chemistry, applications. M.: Nauka, **1987,** [in Russian].
20. Rozantsev, E. G.; Goldfein, M. D.; and Pulin, V. F.; Organic Paramagnetic. Saratov: Saratov Government University; **2000**, 340 p. [in Russian]
21. Liechtenstein, G. I.; *J. Mol. Biol.* **1968**, *2*, 234.
22. Konovalova, N. P.; Bogdanov, G. N.; and Miller, V. B.; Abstract of USSR Academy of Sciences, T. **1964**, *157(3),* 707 p. [in Russian]
23. Suskina V. I.; Candidate dissertation. M.: Academy of Sciences ICP USSR, **1970,** [in Russian]

CHAPTER 4

EFFECTS OF NANOCLAY ON MECHANICAL PROPERTIES OF AGED ASPHALT MIXTURE

M. ARABANI, A. K. HAGHI, and R. TANZADEH

CONTENTS

4.1 INTRODUCTION

Nanotechnology has opened a new world in nanoscale while civil engineering infrastructure is focused on macroscale. Despite the fact that good pavement is constructed using the existing materials, there are significant applications of nanotechnology to improve the pavement performance. Scientists have anticipated that nanotechnology may provide a great potential in the fields of material design, manufacturing, properties, monitoring, and modeling to advance in asphalt pavement technology [1]. One of the most successful polymers is the styrene–butadiene–styrene block copolymer (SBS), which can retard low-temperature thermal stress cracking and improve resistance to high-temperature rutting [2]. Nanomaterial as a new molecular-level size significantly improves properties at mechanical and related levels. Nanoclay composite in the field of nanotechnology field provides a new insight into fundamental properties of clays [3]. The first polymer–clay nanocomposites (PCN) of nylon 6-clay hybrid was invented in 1985 [4]. A number of bitumen physical properties enhanced successfully when as a polymer was modified with small amount of nanoclay [5]. Styrene–butadiene–styrene triblock copolymer used for modifying physical, mechanical, and rheological properties of bitumen [6]. Asphalt concrete is a mixture of bitumen and aggregates and engineers attempt to improve performance of asphalt pavements. One of the most conventional ways to improve pavement performance is bitumen reinforcement with different additives, but the most popular bitumen modification technique is polymer modification [7–9]. In this study, bitumen was aged using TFOT and modified by different percents of nanoclay composite. Marshal samples were made by nanocomposite modified aged bitumen and put under static loading. The results showed improvement in creep parameters and according to images of scanning electron microscope, nanoclay composite was uniformly distributed in the bitumen.

4.2 MATERIALS AND TEST PROGRAM

4.2.1 MATERIALS

According to 101 Iranian Issue [10], middle range of dense graded HMA in Topeka and Binder was used, the specifications of which are given in Table 4.1.

TABLE 4.1 The gradation of the aggregates used in this research

Sieve Size	Passing Percentage
	Topeka and Binder
1 inch	–
¾ inch	100
½ inch	80
3/8 inch	–
#4	59
# 8	43
#50	13
#200	6

The bitumen used for this study was kindly supplied by courtesy of refinery in Tehran. The polymer SBS with specified 5 percent with different nanoclay percent was used. Engineering properties of the bitumen, the nanoclay and SBS are presented in Tables 4.2, 4.3, and 4.4, respectively.

TABLE 4.2 Properties of the bitumen used in this research

Specific Gravity (at 25°C)	Penetration Grade (mm/10)	Softening Point (°C)	Viscosity Pa s	Ductility (cm)	Flash Point (°C)	Purity Grade (%)
1.013	67	50	289	112	308	99.6

TABLE 4.3 Properties of the nanoclay used in this research

PM	Appearance	Purity grade (%)	SSA (m^2/cm^3)	APS (nm)
Nearly spherical	White powder	99	750	10–25

TABLE 4.4 Properties of the SBS used in this research

Polymer	Type	Styren Percent	Molecular Structure
SBS	Carl Prene 501	31	Linear

4.2.2 SAMPLE PREPARATION

To simulate short-term aging of bitumen with respect to the existing facilities, initial bitumen in the laboratory of Guilan University was aged using thin film oven test (TFOT). In this case, bitumen was heated for 5 h at 163°C. Afterward, the obtained aged bitumen used for Marshal samples. In the Marshal samples, construction of reinforce aged bitumen with nanoclay composite, first, nanoclay composite percentage with 2,4,6 and 8 percent bitumen weight was added to the asphalt mixture. The modified bitumen was mixed with different percentages of nanoclay composite for 20 min at temperatures varying from 120 to 150° at 28,000 rpm; this mixer was built by the authors of this article. Then, Marshal mixed design was carried out for both modified and unmodified mixtures.

4.2.3 STATIC CREEP TEST

This test determined the resistance to permanent deformation of asphalt mixtures at temperatures and loads similar to those experienced by these mixtures in the actual field. Creep properties including stiffness, permanent strain, and slop could be determined. Nottingham device test (NDT) was used for not to destroy experimental tests. In this study, stress was considered as 150 KPa and the test was performed at constant temperature of 40°C.

4.2.4 SCANNING ELECTRON MICROSCOPE

Observation of the bitumen structure revealed that bitumen could be also described as a complex binder with two important parts, in which that asphaltenes as a nanometers solid part was surrounded with by an oily liquid matrix (maltenes) [11]. To investigate nanoclay composite effect on bitumen structure and evaluate bitumen distribution, scanning electron microscope (SEM) was used. SEM produces images by scanning a sample using a focused beam of electrons.

4.2.5 FINE PARTICLE HOMOGENEOUS SCATTER

Uniform nanocomposite distributing in the bitumen is an important factor for obtaining stablebitumen [12]. When nanocomposite particles are accumulated, bitumen nanostructure would change the bitumen behavior. With respect to the existing facilities, a new device, as shown in Figure 4.1, was made. A microingredient blender was also made to uniformly distribute nanocomposite-ingredients in the bitumen considering the activities being executed.

FIGURE 4.1 Fine particle homogeneous scatter.

4.3 RESULTS AND DISCUSSIONS

4.3.1 MARSHAL TEST RESULTS

To determine optimum bitumen percent, Marshal samples were made according to ASTM D6926. With these tests, it was found that different nanoclay composite percentages did not significantly affect optimum bitumen value. Thus, to observe the overall evaluation of changes, Marshal test diagrams of aged sample and modified aged sample with 2percent nanoclay composite are given in Figure 4.2.

Results of the Marshall experiments were used on aged and modified nanoclay composite ingredients to determine percentage of optimized bitumen. As an example, graphs of the modified bitumen using 2 percent nanoclay composite were compared with the aged bitumen. Special weight and stability of bitumen samples modified by nanoclay composites increased owing to the existence of nanoingredients and increase in bitumen viscosity. Considering the calculation of actual special weight of the asphaltic mixture, by increasing nanoingredient, not only sample weight of air would increase but also saturated samples' weight with dry level in air would decrease owing to filling of empty spaces of asphaltic mixtures. Using nanoclay composite ingredients and lack of water penetration to the asphaltic mixtures and according to the results of Marshal experiments, special weight of asphaltic mixture was increased to a certain level. This value remained almost unchanged by increasing percentage of nanoingredients. By increasing nano composite ingredients' weight of dry asphalt in air, decreasing sample weight of asphalt in water after being put under vacuum pressure and increasing actual special weight of asphalt, empty space percentage of asphaltic samples was compressed and diminished; by increasing percentage of nanocomposite ingredients, this value did not dramatically change. Marshal stability is maximum resistance against deformation which can be calculated for a constant loading rate. Value of Marshal stability changes with type and gradation of aggregates and type, amount, and calibration of bitumen. Different institutions have various criteria for Marshal stability. In fact, stability of asphaltic mixture is the maximum force that can be tolerated by the sample right before it breaks [13]. The samples are loaded by loading jacks or loading machines with constant rate of 50 ± 5 per minute until the load indicator gauge shows decreased in load rate. The maximum applied load by the machine or the maximum force calculated from the conversion of maximum gauge record is considered to be Marshal stability. According to Marshal experiment, by adding nanoclay composite to old bitumen, actual special weight of the compressed sample increased and special weight of the aggregates mixture decreased; therefore, empty space of the aggregates decreased. By adding more percentage of nanoingredients, these changes continued. Marshal flow is determination (elastic to extra plastic) of asphaltic mixture attained during the experiment. If rate of flow in the chosen optimized bitumen was higher than the upper limit, the mixture would become too plastic and would be considered instable; if it was less than the lower limit, it would be considered too fragile (upper and lower limits were attained according to Iranian Pavement Regulations of 234 Publication). In other words, Marshal flow which was obtained using the flow measurement was deformation of the whole sample between the point at which the load was not applied and the point at which the maximum load started to diminish. According to Marshal experiment, by adding nanoclay composite to old bitumen owing to the increase in bitumen volume, the empty space filled with bitumen would increases. By adding more percentage of nanoingredients, these changes would continue. Optimum bitumen percent changed from 5.51 to 5.98 by adding 8 percent nanoclay composite to aged bitumen.

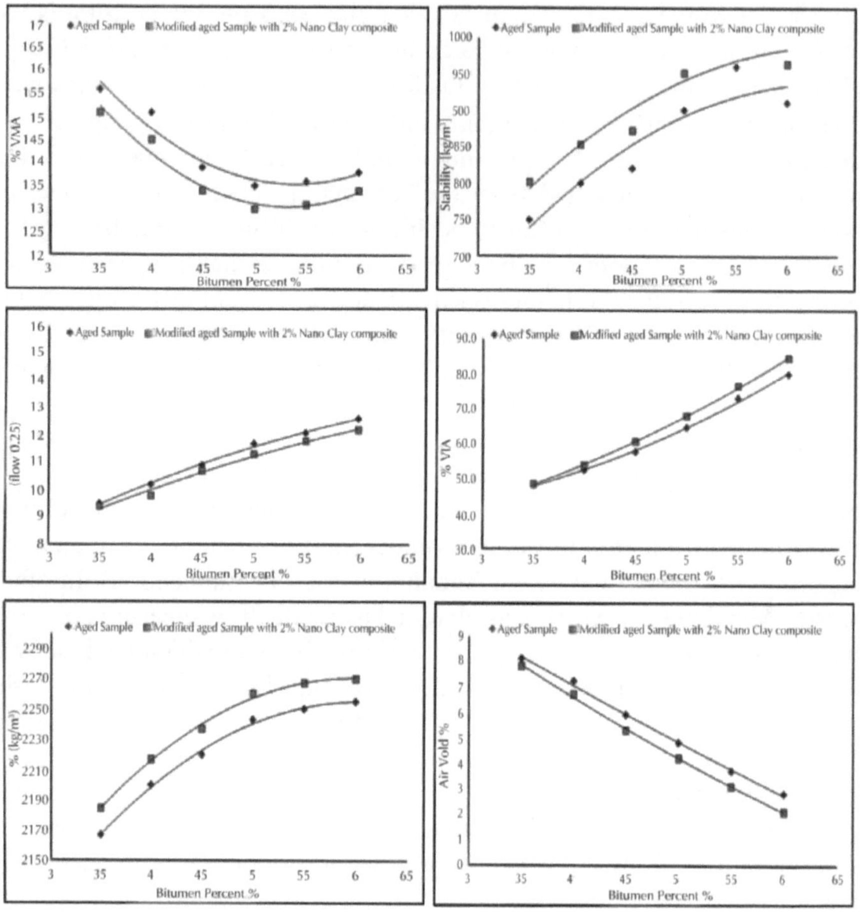

FIGURE 4.2 Diagrams of Marshal test for aged sample and modified aged sample with 2 percent nanoclay composite.

4.3.2 STATIC CREEP ON MODIFIED AGED ASPHALT SAMPLES

Results of static creep tests results are determined as shown in Figure 4.3. These tests measure a specimen's permanent deformation and are conducted by applying a static load to modified aged bitumen and then measuring the specimen's permanent deformation after unloading. High rutting potential is associated with a large amount of permanent deformation. In this study, stress of 150 KPa and temperature of 400°C determined. Creep loading was placed for 1,000 s and then unloading was carried out to 200 s according to Figures 4.3, 4.4, 4.5 and 4.6.

As can be observed, static creep of the asphaltic mixture decreased as percentage of clay nanoingredients added to the bitumen increased; this process continued to the 4 increase. After this value, it diminished owing to the inappropriate connection between bitumen and aggregate caused by accumulation of nano composite–ingredients. Considering the above results, it is clarified that total and permanent strain decreased after adding nanoclay composite in value of 4 percent because nanoclay composites had the maximum value. After adding 4 percent nanoclay composite ingredients, asphalt mixture's strain increased owing to shriveling of the bitumen fall in its flexibility. For better view about the strain values, Figure 4.4 shows in the logarithmic plot.

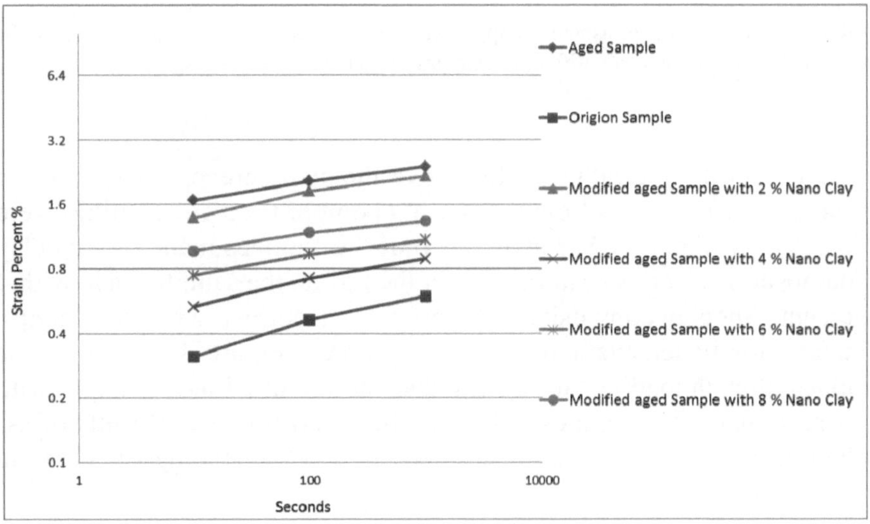

FIGURE 4.3 Static creep test on modified aged samples with different nanoclay composite

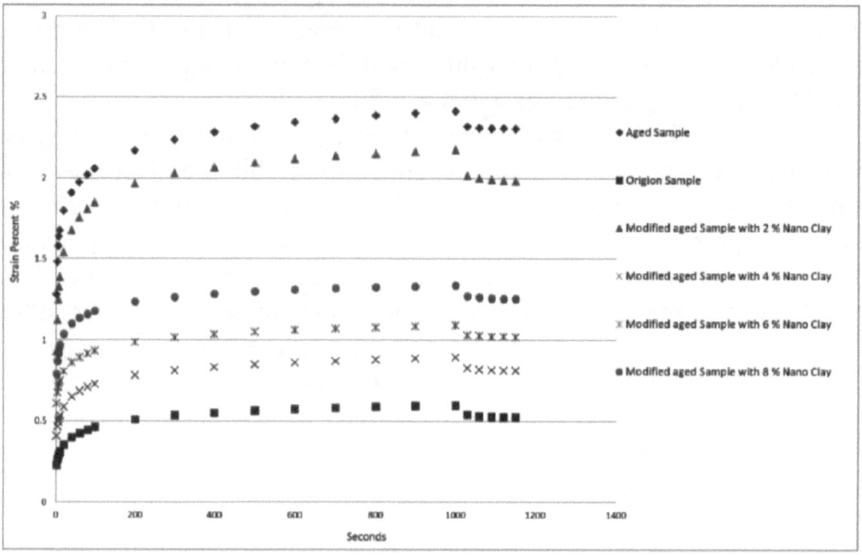

FIGURE 4.4 Results of asphalt samples' Static Creep test at 10, 100, and 1000 s on the modified aged samples with different nanoclay composite percentages.

Also, in the logarithmic plot above the static creep, changes in the time of 10,100, 1000 s from the starting point of the static loadings were depicted. In Figure 4.5, elastic creep of the samples made of nanoclay composite is clear. As can be seen in the figure, the sample made with 4 percent nanoclay composite had more ramp (returned with more ramp). In fact, for better evaluation of elastic creep, Figure 4.5 shows these values after unloading. It is clear that the modified aged sample with 4percent nanoclay composite had the best performance. Results of asphalt samples' permanent static creep test after unloading are given in Figure 4.6.

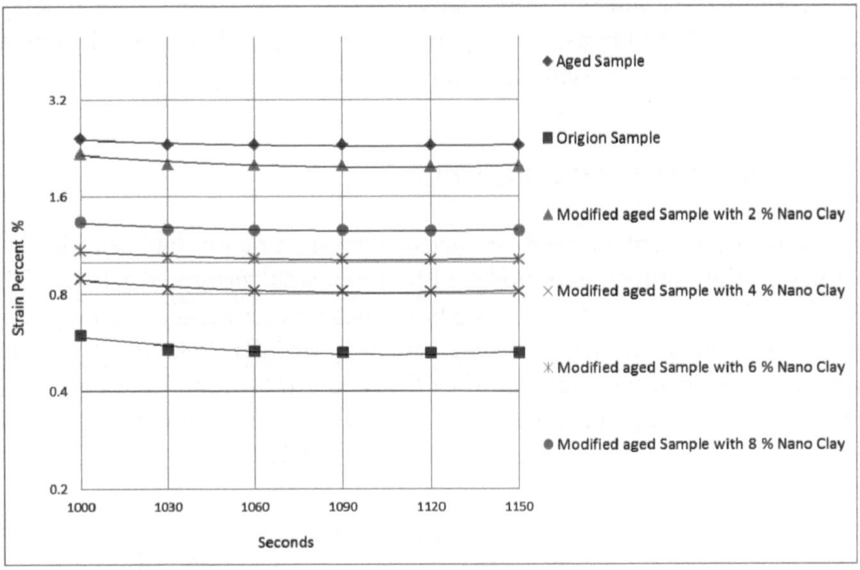

FIGURE 4.5 Asphalt samples' elastic static creep test after unloading on the modified aged samples with different nanoclay composite percentages.

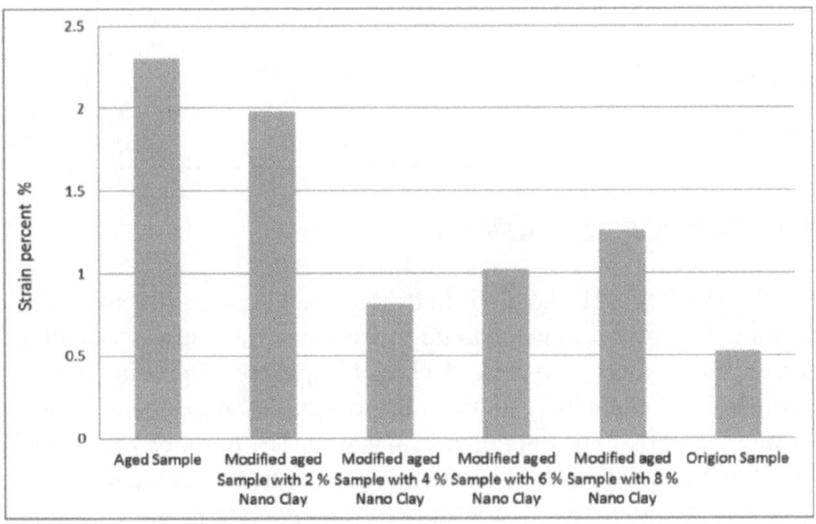

FIGURE 4.6 Asphalt samples' permanent static creep test after unloading.

According to Figure 4.6, modified aged sample with 4 percent nanoclay composite, had the lowest strain percentage compared with aged sample and asphalt strain increased after adding nanoclay composite.

4.3.3 MODULUS TEST RESULTS

Modulus experiment on the resilience of the asphaltic mixture samples including initial bitumen and modified bitumen was performed 50°C. With the ASTM-D4123 standard methods in these experiments, loading pattern of semisinusoidal with loading frequency of 1 Hz, loading cycle time period of 1 s, loading time period of 0.1 s, and relaxation time period of 0.9 s in each loading cycle were used. Five pulses for determining asphalt sample's resilient modulus are shown in Table 4.5.

TABLE 4.5 Resilient modulus results of asphalt samples

Aged Samples	**Pals1**	**Pals2**	**Pals3**	**Pals4**	**Pals5**	**Average Modulus (MPa)**
	230	242	222	233	265	238/4
Modified agedsample with 4 % Nanoclay composite	Pals1	Pals2	Pals3	Pals4	Pals5	Average Modulus (MPa)
	248	261	243	241	284	255/4

4.3.4 FATIGUE TEST RESULTS

According to the results taken from the Nottingham experiment and by determining the optimized percentage of nanomaterials, it was clarified that 4 percent nanoclay composite had a better impact on improving. Functional properties of asphaltic mixtures, these samples were made under 150 KPa stress at 25°C to be put under exhaustion. The purpose was to analyze exhaustion operation of the samples which had the best creep operation. In Figure 4.7, the modified aged samples with 4 percent nanoclay composite compared with aged samples had higher strain against the aged samples in the nearly same loading cycles. These results confirmed that nanoclay composite did not have a very good effect on fatigue performance [14].

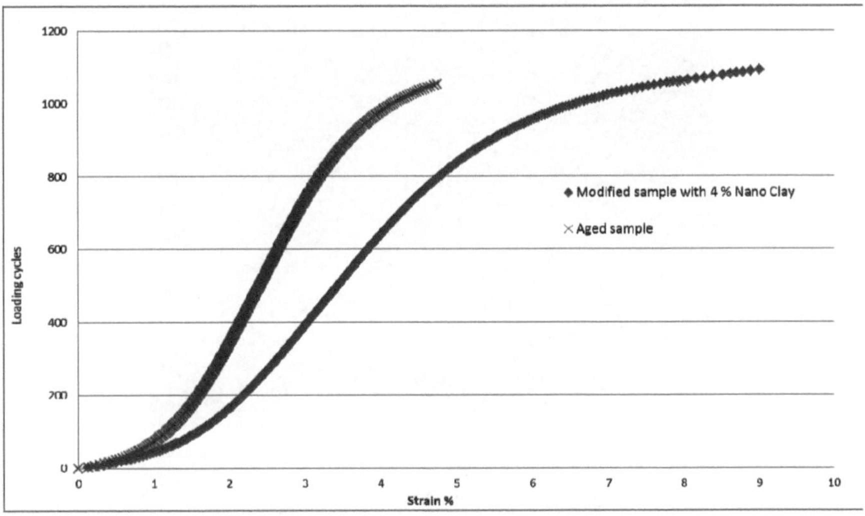

FIGURE 4.7 Fatigue test results of asphalt samples on the modified aged samples with 4 percent nanoclay composite percent and aged sample.

4.3.5 SEM RESULTS

About 20 years ago, this microscope was invented and scientists observed atoms. This microscope works through assessing very small flows made between the microscope tip and the sample. The wire charged with positive loads by the thin tip in the microscope which is in the distance of about one nanometer from surface of the object causes a weak flow and throw of the surface electron to the detector (microscope wire). These images were taken from the bitumen in the thin layer laboratory in Iran University of Science and Technology. Because of high thermal sensitivity of bitumen and its ruining probability in the case of increased system voltage, quality of the image was not so good.

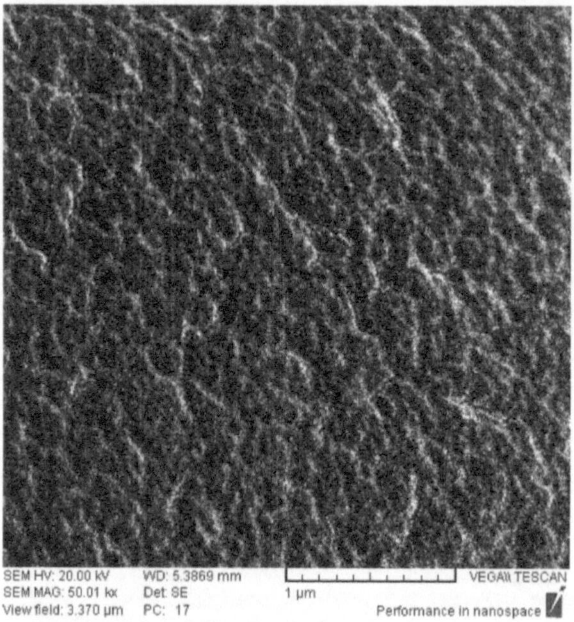

FIGURE 4.8 Aged bitumen image with SEM.

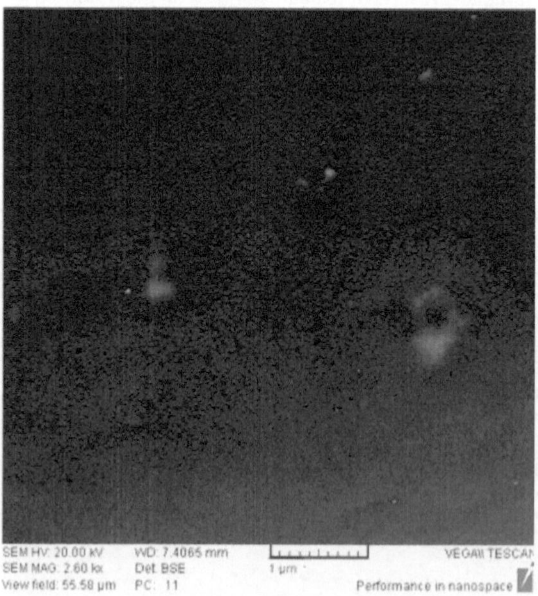

FIGURE 4.9 Modified aged bitumen with nanoclay composite image with SEM.

The nanocomposite–ingredients used in this research were almost spherical in shape, which is shown in the figures related to the ingredient. SEM images for identifying the size of nanocomposite can be observed in Figures 4.8 and 4.9. Nanocomposite's uniform distribution can be also notified in bitumen. As seen in these images, mechanical mixing method had good agreement with the innovative device developed in this article.

4.4 CONCLUSIONS

In this study, bitumen was aged using TFOT and then modified by different nanocomposite percentages. Marshal samples were made using modified aged bitumen with different nanocomposite percentages and aggregates with Topeka and Binder gradation. These samples were put under static creep loading and those with high creep resistance were placed under fatigue loading and defined resilient modulus. According to the conventional tests, adding 4 percent nanocomposite in bitumen returned the missed properties of bitumen by strengthening the bond between bitumen components. After a few years, bonding between aggregates and bitumen decreased and, with creating a gap, aged asphalt mixture's creep increased. In this study, static creep test results on the asphalt mixtures made by modified aged bitumen were observed. Nanoclay composite decreased total strain of aged asphalt mixtures made by modified bitumen and, according to creep test results, 4 percent nanoclay composite was a decent amount. This research was one of the pioneers on the asphaltic samples made of aged bitumen and aged bitumen modified by nanocomposite ingredients of clay. Generally, nanoingredients decrease thermal sensitivity of bitumen and the whole mixture. Asphalt samples' resistance against static loading also increased. Resilient modulus increased, and it seems that nanoclay composite ingredients did not have a positive impact on the mixture operation under the fatigue load.

REFERENCES

1. Steyn Wyn and Jvd M.; Applications of Nanotechnology in Road Pavement Engineering, Springer-Verlag Berlin Heidelberg, Nanotechnology in Civil Infrastructure, **2011**, 49–83.

2. Sabbagh, A. B.; and Lesser, A. J.; Effect of particle morphology on the emulsion stability and mechanical performance of polyolefin modified asphalts. *Poly. Eng. Sci.* **1998**, *38(5),* 707–715.
3. Chong, K. P.; Nanotechnology in civil engineering. In: Proceedings of 1st International Symposium on Nanotechnology in Construction. Paisley, Scotland, **2003**, 13–21.
4. Gang Liu; Martin van de Ven; and Andre Molenaar; Influence of organo-montmorillonites on fatigue properties of bitumen and morta, journal homepage: www.elsevier.com/locate/ijfatigue. *Int. J. Fatig*. **2011**, *33*, 1574–1582.
5. Saeed Ghaffarpour Jahromi; and Ali Khodaii; Effects of nanoclay on rheological properties of bitumen binder, journal homepage: www.elsevier.com/locate/conbuildmat. *Construc. Build. Mater.* **2009**, *23,* 2894–2904.
6. Saeed Sadeghpour Galooyak; Bahram Dabir; Ali Ehsan Nazarbeygi; and Alireza Moeini; Rheological properties and storage stability of Bitumen/SBS/montmorillonite composite, journal homepage: www.elsevier.com/locate/conbuildmat. *Construc. Build. Mater.* **2010,** *24*, 300–307.
7. Sayyed Mahdi Abtahi; Mohammad Sheikhzadeh; and Sayyed Mahdi Hejazi; Fiber-reinforced asphalt-concrete–A review. *Construc. Build. Mater*. **2010**, *24*, 871–877.
8. Shen, S.; Amirkhanian, J.; and Miller, A.; Effects of rejuvenator on performance-based properties of rejuvenated asphalt binder and mixtures. *Constr. Build. Mater.* **2007**, *21(5),* 958–964.
9. Haghi, A. K.; Arabani, M.; Shaker, M. i; Haj jafari, M.; and Mobasher, B.; Strength modification of Asphalt pavement using waste tires. In: 7th International Fracture Conference, University of Kocaeli, Kocaeli, Turkey, **2005**.
10. 101 Iranian Issue. General Technical specification of Roads, http://tec.mporg.ir, **2010**.
11. Zhanping You; Julian Mills-Beale; Justin M. Foley; Samit Roy; Gregory M. Odegard; Qingli Dai; and Shu Wei Goh; Nanoclay-modified asphalt materials: Preparation and characterization. *Construc. Build. Mater*. **2011,** *25*, 1072–1078.
12. Jian-Ying Yu; Peng-Cheng Feng; Heng-Long Zhang; and Shao-Peng Wu; Effect of organo-montmorillonite on aging properties of asphalt. *Constr. Build. Mater*. **2009**, *23,* 2636–2640.
13. ASTM D6927. Standard Test Method for Marshall Stability and Flow of Bituminous Mixtures. **2007**.
14. Van de Ven, M. F. C.; and Molenaar, A. A. A.; Nano Clay for Binder Modification of Asphalt Mixtures. London: Taylor & Francis Group; **2009,** ISBN 978-0-415-55854-9

CHAPTER 5

A LECTURE NOTE ON QUANTUM CHEMICAL CALCULATION STUDIES THE MECHANISM OF PROTONATION OF 2-ETHYLBUTENE-1 BY METHOD DFT

V. A. BABKIN, D. S. ANDREEV, G. E. ZAIKOV, and G. K. ROSSIEVA

CONTENTS

5.1 AIMS AND BACKGROUNDS

According to modern conceptions about the mechanism of initialization of cationic polymerization of 2-ethylbutene-1 the true catalyst of this reaction is Lewis' aqua acids ofthe type $AlCl_3{\times}H_2O$, $AlCl_2C_2H_5{\times}H_2O$, $BF_3{\times}H_2O$ and others (i.e., there are always admixtures of water in the system) out of which due to complex coordinated interactions initiating particle $H^{+\delta}$ forms and which is in turn according to Morkovnikov's rule attacks the most hydrogenated atom of carbon C_α [1–3]. Studying the mechanism of 2-ethylbutene-1 protonation is the first step in studying of the mechanism of elementary act of initialization of cationic polymerization of this monomer.

In this connection the aim of the present work is quantumty–chemical study of the mechanism of 2-ethylbutene-1 protonation by classical method **D**ensity **F**unctional **T**heory (DFT).

5.2 METHODOLOGY

For studying the mechanism of protonation the classical quantum–chemical method DFT-PBE0/3-21g was chosen with geometry optimization of all the parameters by gradient method built in PC GAMESS [4], as this method specifically parameterized for the best reproduction of the energy characteristics of molecular systems and it is an important factor in analysis of the mechanisms of cation processes. Calculations were done in the approximation of an isolated molecular in a gas phase. In the system $H^+ \ldots C_6H_{12}$ (2-ethylbutene-1) 19 atoms, M=2S+1=1 [where S is the total spin of all electrons of study system is zero (all electrons are paired), M is the multiplicity], the total charge of a molecular system $\Sigma\, q_c = 1$. It was done the calculation of potential energy of proton interaction with 2-ethylbutene-1 for studying the mechanism of2-ethylbutene-1 protonation by the following way. The distances from proton H_1 up to C_2 (R_{H1C2}) and from H_1 up to C_3 (R_{H1C3}) were chosen as reaction coordinates. The original meanings of R_{H1C2} and R_{H1C3} were taken as 3.1 Å. Further, changing meanings of $R_{H1C2}$with 0.1 Å step quantum–chemical calculation of molecular system was done changing R_{H1C2} meanings with the same 0.1 Å step. According to the received data of energy meanings equipotential surface of proton interaction with 2-ethylbutene-1 was built along the reaction coordinates (see Figure 5.4). The initial model of the proton attack of 2-ethylbutene-1 molecule is shown in Figure 5.1.

The famous program MacMolPlt [5] was used for the visual representation of molecules' models.

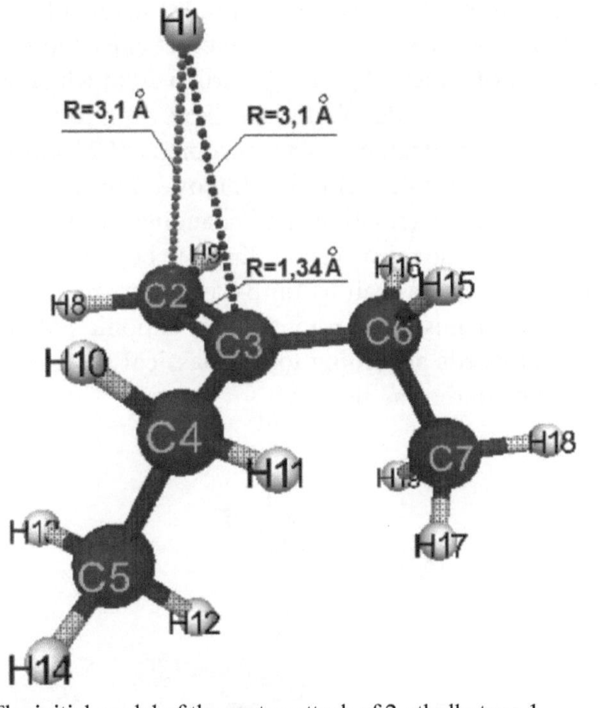

FIGURE 5.1 The initial model of the proton attack of 2-ethylbutene-1.

5.3 RESULTSAND DISCUSSION

The energies of the molecular system $H^+ \ldots C_6H_{12}$along the reaction coordinates R_{H1C2} and R_{H1C3} were shown in Table 5.1. The final structure of the formed carbcation after proton H_1 attack of α-carbon atom of 2-ethylbutene-1 (C_2) and a break of the double bond of 2-ethylbutene-1 is shown in Figure 5.2. The final structure of the formed carbcationafter proton H_1 attack ofβ-carbon atom of 2-ethylbutene-1 (C_3) and a break of the double-bond $C_2 = C_3$is shown in Figure 5.3. The charges on atoms of the final structures of formed carbcations are introduced in Table 5.2. The changing of the total energy under protonization of 2-ethylbutene-1 is shown in Figure 5.4, it is seen that the initiating particle $H^{+\delta}$ along the reaction coordinates R_{H1C2} and R_{H1C3} negative value of the total energy of the system $H^+ \ldots C_6H_{12}$ (E_0) is steadily increasing up to the complete formation of carbcation (see Figure 5.4) on the whole way of proton movement having barrier-free nature as well as under the attack of proton on α- and β-carbon atoms of 2-ethylbutene-1. However, the final structure of the attack of proton on α carbon atom per 53 kJ/mol is energetically more favorable than the final structure of attack of proton β-carbon atom which is in full accordance with the classical rule of Morkovnikov. As a result of this reaction the energy gain under attack on α-carbon atom is 509 kJ/mol and under attack on β-carbon atom is 456 kJ/mol. Moreover, the analysis of the results of quantum–chemical calculations and changing of the bond lengths and valence angles along the reaction coordinate in both cases under the attack of proton on α-carbon as well as on β-carbon atoms of 2-ethylbutene-1 testify that the mechanism of protonation of cationic polymerization of2-ethylbutene-1 proceeds according to the classical scheme of joining proton to the double bond of monomer.

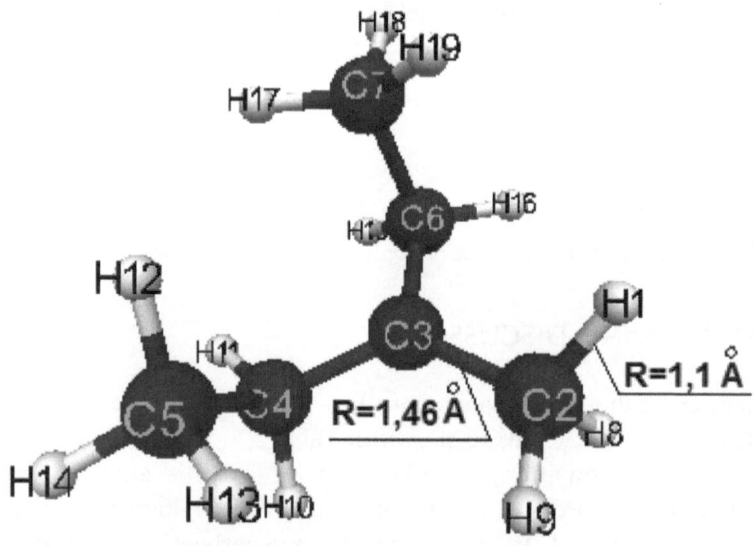

FIGURE 5.2 The final structure of the formed carbcationafter proton H_1 attack of α-carbon atom of 2-ethylbutene-1 (C_2).

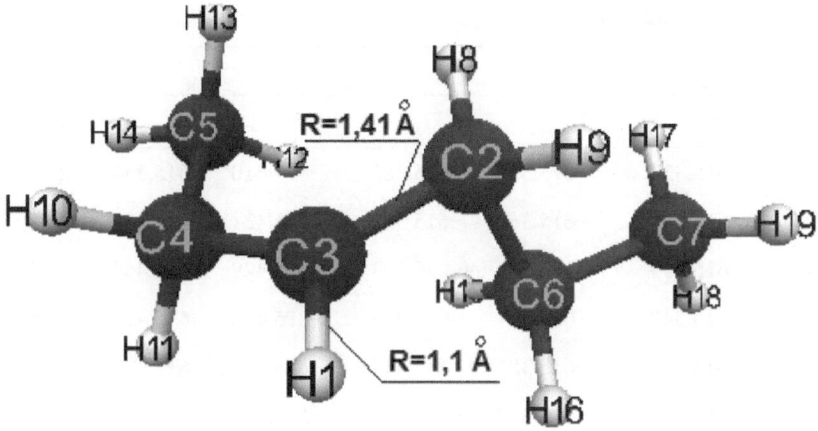

FIGURE 5.3 The final structure of the formed carbcation after proton H_1 attack of β-carbon atom of 2-ethylbutene-1 (C_3).

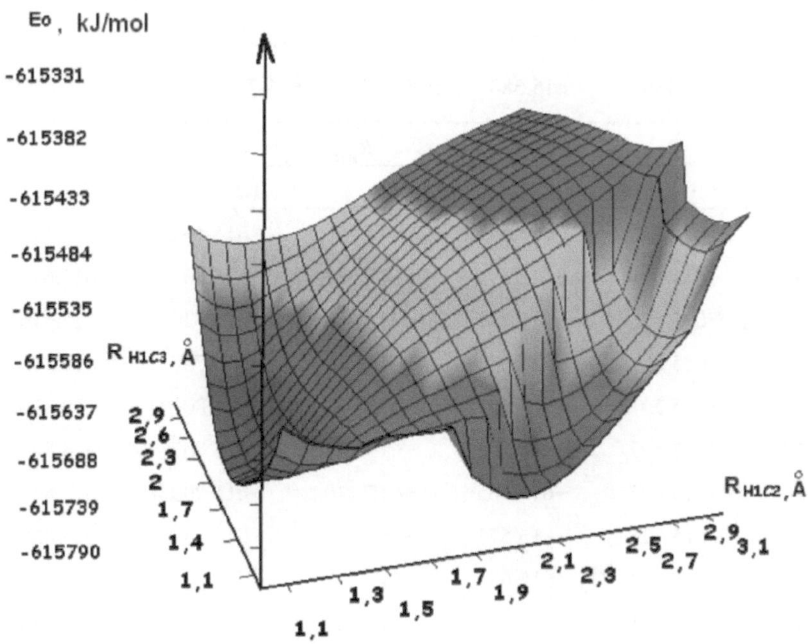

FIGURE 5.4 Potential surface of integration energy of proton with 2-ethylbutene-1 (see Table 5.1).

TABLE 5.1 Values of energy of the molecular system $H^+ \ldots C_6H_{12}$ - E_o (inkJ/mol) along the reaction coordinates R_{H1C2} and R_{H1C3}(inÅ)

R_{H1C3}	R_{H1C2}					
	3.1	**2.9**	**2.7**	**2.5**	**2.3**	**2.1**
3.1	−615.332	−615.355	−615.381	−615.410	−615.444	−615.484
2.9	−615.337	−615.360	−615.387	−615.421	−615.458	−615.502
2.7	−615.339	−615.366	−615.394	−615.429	−615.468	−615.515
2.5	−615.347	−615.374	−615.402	−615.436	−615.476	−615.526
2.3	−615.355	−615.381	−615.413	−615.447	−615.486	−615.534
2.1	−615.363	−615.397	−615.429	−615.463	−615.500	−615.544
1.9	−615.374	−615.418	−615.452	−615.486	−615.520	−615.560
1.7	−615.374	−615.542	−615.484	−615.520	−615.552	−615,586
1.5	−615.439	−615.589	−615.636	−615.654	−615.597	−615.623
1.3	−615.458	−615.607	−615.678	−615.728	−615.741	−615.665
1.1	−615.439	−615.584	−615.657	−615.730	−615.778	**−615.788**

R_{H1C3}	R_{H1C2}				
	1.9	**1.7**	**1.5**	**1.3**	**1.1**
3.1	−615.520	−615.547	−615.560	−615.562	−615.536
2.9	−615.549	−615.599	−615.636	−615.660	−615.631
2.7	−615.570	−615.631	−615.686	−615.730	−615.715
2.5	−615.581	−615.646	−615.717	−615.780	−615.786
2.3	−615.591	−615.657	−615.730	−615.801	−615,830
2.1	−615.597	−615.662	−615.733	−615.807	**−615.841**
1.9	−615.607	−615.665	−615.730	−615.799	−615.833
1.7	−615.626	−615.673	−615.725	−615.783	−615.809
1.5	−615.652	−615.683	−615.723	−615.767	−615.778
1.3	−615.681	−615.699	−615.720	−615.744	−615.736
1.1	−615.749	−615.688	−615.688	−615.694	−615,665

TABLE 5.2 Charges of atoms of the final models of formed carbcations

Atom	Charges on Atoms of Formed Carbcation	
	After the Attack of H_1 Proton α-Carbon Atom of 2-Ethylbutene-1 (C_2)	After the Attack of H_1 Proton β-Carbon Atom of 2-Ethylbutene-1 (C_3)
H(1)	+0.32	+0,34
C(2)	−0.71	−0.55
C(3)	+0.22	+0.03
C(4)	−0.55	−0.58
C(5)	−0.64	−0.65
C(6)	−0.55	−0.49
C(7)	−0.65	−0.63
H(8)	+0.34	+0.32
H(9)	+0.30	+0.33
H(10)	+0.33	**+0.35**
H(11)	+0.29	+0.34
H(12)	+0.27	+0.25
H(13)	+0.26	+0.26
H(14)	+0.30	+0.29
H(15)	+0.32	+0.29
H(16)	+0.30	+0.29
H(17)	+0.27	+0.26
H(18)	+0.31	+0.29
H(19)	+0.27	+0.26

5.4 CONCLUSION

Thus we have studied the mechanism of protonation of 2-ethylbutene-1 by quantum–chemical method DFT for the first time. It is shown that this mechanism is a usual reaction of joining proton to double bond of olefin. The reaction is exothermic and has a barrier-free nature. It is energetically favorable for the reaction to follow the classical scheme in accordance with Markovnikov's rule.

KEYWORDS

- **2-ethylbutene-1, DFT Method**
- **Mechanism of reactions**
- **Protonation**
- **Quantum chemical calculation energy**

REFERENCES

1. Kennedy, J.; Cationic polymerization of olefins. J. Kennedy. M., **1978**, 431 p.
2. Sangalov, Y. A.; Polymers and copolymers of isobutylene. Sangalov Y. A.; and Minsker, K.; S.: Ufa, **2001**, 381 p.
3. Babkin V. A.; Zaikov G. E.; and Minsker K. S.; The quantum-chemical aspect of cationic polymerization of olefins, **1996**, Ufa, 182 p.
4. Shmidt, M. W. J.; *Comput. Chem.* M. W.; Shmidt, M. S.; Gordon (and another). **1993**, *14,* 1347–1363 pp.
5. Bode, B. M. J.; *Mol. Graph. Mod.* B. M. Bode, M. S.; Gordon. **1998**, *6,* 133–138 pp.

CHAPTER 6

INFLUENCE SOLUTIONS OF GLYCEROL ON THE ENZYMATIC ACTIVITY OF PROTEOLYTIC COMPLEX OF HEPATOPANCREAS CRAB STABILIZED POLYSACCHARIDE COMPOUNDS

A. A. BELOV, A. I. KOROTAEVA, and E. A. RASPOPOVA

CONTENTS

6.1 AIM AND BACKGROUND

The important prerequisite to intensive development of a medical enzymology was the solution of theoretical and technological questions of enzyme's modification with the purpose of giving of new properties to them, and such which have cardinal value for therapy of diseases of the person [3]. The consideration of results of chemical modification (and for our work—an immobilization on insoluble carriers) of therapeutic active agents is interested for the search of new medicines of the proteinaceous nature. Now there are the possibilities of creating of fibrous materials possessing the biological activity of the most various range of action [2, 3, 8, 9]. Such products matter not only for medical practice and, especially, operational surgery, but also for various branches of equipment and biology. The connection between polymeric matrix and biologically active substance can be covalent, coordinational, and/or ionic at modification of fibers for the purpose of giving them the biological activity. The choice of each of the called types of communication is defined by practical purpose of created materials. Chemical modification of protein's globule can significantly influence on enzyme's stability. Thus, the development of the available atraumatic dressing containing biologically active agents (enzymes) that allows to reduce terms of treatment and to increase the efficiency of rendered medical actions, represents an actual scientific problem [3, 4].

The hydratation of proteins, that is, the quantity and distribution of water molecules on a biopolymer surface, is one of the major factor influencing on a protein's folding and a formation of their native conformation. Directly or indirectly water participates in all noncovalent interactions stabilizing protein's structure. However, the hydrate water forming adjacent layers to a protein's molecule, first of all is necessary for functioning of proteins, in particular enzymes. Low-molecular organic compounds can act as regulators of enzyme's activity *in vitro* and *in vivo* [10–14]. The influence of organic compounds on protein may be as direct through intermolecular contacts, and mediated owing to the change of structure and dynamics of a hydrate cover [11]. It should be noted that the use of organic solvents for modification of hydrate water structure in proteins systems is methodical approach which allows to influence on structure of a hydrate cover selectively.

One of the problems arising at utilizing of cellulose fibrous applications, is their injury—napkin sticking to a wound after evaporation of moistening solution (preparations with the immobilized enzymes in a dry form are practically inactive). Besides, the surface of a wound is dehydrated [8], drying on air or under dressing and the firm crust (scab) on it which consists of the dried exudates' and cellular detritus is formed. The similar dried layer represents a natural obstacle for migrating epidermocytes that considerably increased healing time. Therefore there arised a problem of creation of the optimum environment for wound healing before developers of dressing means: it is harmful both excess of moisture and its deficiency. Thus, moistening solution is urged to solve at the same time some problems: it must be a bandage less traumatic, to provide the environment for coursing of enzymatic reaction and to create the optimal environment for wound healing. The search of the substances raising an atraumaticness of modified cellulose dressing materials without decreasing of enzymes activity is have continued.

We studied the influence of glycerol (Gl) additives to moistening solution on change of enzymatic activity of the various forms of PC. The Gl is allowed for using in medical practice [1], and there is a large number of literary data about stabilization of proteolytic enzymes by Gl solutions [11–14]. Obtained data showed, that even 2 percent solution of Gl through 72 hours isn't completely evaporate from a surface of a textile material that contributes to preservation of the atraumatic properties of application.

Besides, it was established that various forms of Ch stabilized the PC, protecting their enzymes from denaturizing effects of glycerol.

6.2 INTRODUCTION

The wide experience of the usage of proteolytic enzymes in clinical surgery is so far saved up. It is generalized in many fundamental works of domestic and foreign researchers. Treatment of purulent-necrotic processes pursues the aims of early removal of the devitalized fabrics from a wound, suppression of microflora, and acceleration of regeneration [2, 8]. The local enzymotherapy was firmly strengthened in an arsenal of remedies at treatment is purulent-necrotic wounds. However, despite visible progress, a enzymotherapy as the treatment method, has a number of essential shortcomings. Native proteolytic enzymes are quickly inactivated by inhibitors of blood plasma, they are insufficiently firm in solutions, possess an antigenicity, are exposed to autolysis, are very sensitive to thermal affects and pH changes that leads to significant increasing of preparation's expense in the course of treatment. Besides, enzyme's preparations are expensive and scarce that also considerably limits their application. It was succeeded to overcome these shortcomings by creation of the immobilized enzymes [1–5]. The important prerequisite to intensive development of a medical enzymology was the solution of theoretical and technological questions of enzyme's modification with giving of new properties to them, and such which are important for therapy of diseases of the person [3].

The modification (immobilization) of enzyme (capable selectively to limit its contact with cell receptors), allows to solve any more problem. It is the most real way for restriction of damaging action of this group of preparations [3, 5].

The microfibrous structure of biocompatible and biodegradative polymers, containing the medical preparations which were released in process of their contact with a wound surface [4] is satisfied to the majority of requirements imposed to the wound and burn dressing materials (DM). It is possible to receive materials with the fillers evenly or superficially distributed in fiber depending on the purpose of a product. The dosed delivery of drugs means the destruction of the matrix containing therapeutic components. One of the perspective biopolymers possessing ability to biodegradation is nontoxic chitosan [5]. Thus, it is necessary to develop the suitable structure of dressing materials and technology of its receiving for creation of DM meeting the requirements of polyfunctionality. The aforementioned is impossible without comprehensive investigation of physical and chemical characteristics of polymeric solutions and an analysis of the influence of technological parameters on properties of received fibrous structures. Thus, development of the available atraumatic dressing materials, allowing to reduce terms of treatment and to increase efficiency of rendered medical actions, represents an actual scientific problem [3, 4].

Since ancient times and to this day natural cellulose fibers, such as cotton and flax are used in medicine. Moreover, cellulose fibrous materials represent one which cannot be replaced on any other.

Fibrous (textile) materials, which are used in the medical field or health services, can be divided into two basic groups, according to whether they are used: (1) inside organic tissues (internal/implantable): vascular grafts, meshes, stents, tendons, and ligament implants, surgical threads, etc. or 2) on their surface (external/nonimplantable): gauzes, bandages, surgical covers, nappies, tampons, etc. The use of natural cellulose fibers, such as cotton and flax, goes back in medical applications to ancient times and still today, in some medical applications, cellulose fibrous materials represent materials that can not be exchanged with any other. In the more recent past, new procedures and technologies enabled the production of various chemical cellulose fibers such as viscose, modal, and lyocell, which are cleaner and even more hygroscopic than cotton, and as such highly applicable within hygiene and medical fields [9].

Fibrous carriers of enzymes were very effective when there is a need for the materials possessing along with enzymatic activity good draining properties, i.e. ability quickly to clear a wound surface [2, 8]. Acceleration of terms of healing of wounds, decrease in a consumption of enzymes and dressings [3] belongs to advantages of their use.

High potential of cellulose fibers is connected with their molecular structure which offers excellent opportunities as a matrix for creation of bioactive, biocompatible, and intellectual materials [9]. Natural cellulose is characterized by the insignificant maintenance of carbonyl groups. Ketonic and carboxyl groups in it practically are absent. However, when receiving technical cellulose from vegetable raw materials in processes of cooking and a bleaching the number trailer the aldehyde of groups increases, and there are not trailer aldehyde, and also ketonic and carboxyl groups. It is caused by hydrolytic and oxidizing destruction of macromolecules of cellulose and oxidation of spirit groups [9, 15].

Sharing of fibrous materials and biocompatible gels [e.g., on a basis of chitosan (Ch)] allows to receive the difficult composite materials containing on one carrier, various classes of medicinal substances isolated from each other: enzymes (proteinases) and proteins (insulin); enzymes and antibiotics or antioxidants; enzymes and inhibitors.

At production of wound coverings, cosmetic products, and similar medical dressing materials, it is necessary to take into account that unlike bioactive materials of other appointment these products are disposable remedies with the short term of operation (no more than 72 h, and more often till 24 h), therefore their biological activity has to be implemented as much as possible when imposing on a wound or skin [3].

One of the problems arising at utilizing of cellulose fibrous applications, is their injury—napkin sticking to a wound after evaporation of moistening solution (preparations with the immobilized enzymes in a dry form are practically inactive). Moistening solution is urged to solve at the same time some problems: it must do a bandage less traumatic, to provide the environment for course of enzymatic reaction and to create the optimum environment for wound healing. It is necessary to use as components of moistening solution only the substances allowed for clinical employment. It was established earlier in our work [10] that DMSO and its water solutions don't reduce enzyme's activity of the developed preparations and promote preservation of "the damp environment" on the carrier moistened with it that facilitates removal of a bandage from a wound. But high concentration of DMSO cause burning and itch of skin of the patient. Therefore, search of the substances raising an atraumaticness of our dressing materials without decreasing of enzyme's activity, is proceeding.

Glycerol (Gl) is allowed for use in medical practice [1]; besides, there is a large number of literary data on stabilization by solutions of Gl of proteolytic enzymes [11–13].

There is, however, one factor limiting the wide use of organic solvents and enzymes in medicine. The fact is that the enzymes preserve their unique catalytic properties only in a rather narrow range of conditions, namely in aqueous solutions at neutral or near-neutral pH and at not too high temperatures. However, the balance of important reactions shifted in these conditions, in the direction of the initial reagents. In other words, during using biocatalysts in medicine often a situation arises when the necessary conditions for the preservation of catalytic properties of the enzyme, are incompatible with the thermodynamic conditions optimal for achieving high output catalyzed reaction. The contradiction between the relatively "tough" conditions, in which chemical reaction must proceed, and the "tender" nature of the biocatalyst (relatively labile structure) may be in principle solved. One of the ways of the enzyme's defense against inactivating effects of high temperature, extreme pH values, organic solvents, etc., is their stabilization by immobilization. On this way notable success already achieved, but the researches are continued [3, 14].

6.3 EXPERIMENTAL PART

In this work were used: PC (collagenase for the food industry TU (technical conditions) 9281-004-11734126-00) of production of NPO «Bioprogress» (Shchelkovo, Moscow region, Russia). Chitosan of production of NPO "Bioprogress" (Schelkovo, Moscow region, Russia) humidity drug is 10%, TU (technical conditions) 9289-067-00472124-03, the degree of deacylation, 80%; kinematic viscosity, not less 383.7 cSt; (molecular mass, 478 kDa). All other reagents, unless otherwise noted, domestic production, and qualification not less than "chemically pure.".

Activation of cellulose carrier in the form of woven cloths (medical gauze) realized by periodate sodium, resulting received dialdehydecellulose (DAC) the required degree of modification of secondary alcohol groups [3, 9, 15]. The number of aldehyde groups on carrier was determined analogously to [16] by oxidation of the last solution by iodine in alkaline medium (0.1 H solution $Na_2B_4O_7$) and expressed in mM/g taking into account humidity of carrier.

Immobilization of PC on the cellulose carrier was realized similarly [3, 17]. The maintenance of Ch on the carrier was 8 mg/g carrier, the concentration of the PC was amounted as 8 mg/g carrier. Enzymatic activity of preparations was determined with using azocoll [18, 19] or casein [18, 20] as substrate.

The modification of PC by Ch-solution was carried out as follows: equal volumes of PC (solution 2 mg/ml 0.1 M NaCl) and acetic solution of Ch (to 0.5 mass%) were poured; the resulting gel contained 1 mg of PC and 2.5 mg of Ch in 1 ml of gel.

To determine the inactivation of organic solvent, anhydrous solvent content of the main substance 100 percent pre-washed mixed with 1/15 sodium phosphate buffer (PB) desired pH values (usually 6.2) in a specified ratio (the solution was cooled to room temperature). To the drug-modified PC or the required volume of solution of enzyme, added to the solution of an organic solvent in PB (kept at room temperature for 15 min), in order to determine the residual enzyme activity.

6.4 RESULTS AND DISCUSSION

As it was noted in [8], drying up in the air or under the bandage, surface wounds dehydrated, and a solid crust (scab), which consists of the dried exudates' and cellular debris it formed. Similar dried layer is a natural barrier to migratory epidermocytes, which greatly lengthens the time of healing. So it is necessary to prevent drying of the bandages. We investigated the evaporation of water solutions of Gl from the surface of the fiber-forming polymers. For this purpose matrix in the form of sandwich napkins was soaked with a solution of Gl with setting concentration, pressed and placed in dry-air thermostat (37±1°C, relative humidity 75±15%) at the specified time, the top napkin was covered with a polyethylene film. The kinetics of solvent evaporation from carrier was observed for loss of weight. The obtained data are given in Table 6.1.

In the present work, we did not study a stabilization or destabilization of enzyme systems in the presence of the of glycerol's solutions in dependence on various temperatures that we should investigate for clinical usage of obtained preparations. It is the task of future research.

TABLE 6.1 The weight changing of dressing material's samples during storage after glycerol's processing

WEIGHT WIPES		GL CONTENTS IN THE IMPREGNATING SOLUTION (MASS %)		
		0%	2%	5%
CELLULOSE	BEFORE TREATMENT, G	1,2122	1,1948	1,3432
	AFTER IMPREGNATION, G	3,8046	3,9516	4,2413
	AFTER 24 H, G	1,2140	1,2327	1,4630
	AFTER 48 H, G	1,2086	1,2220	1,4524
	AFTER 72 H, G	1,2086	1,2248	1,4564
DAC (0,4)	BEFORE TREATMENT, G	1,1344	1,1004	1,0926
	AFTER IMPREGNATION, G	3,7681	3,8610	3,6014
	AFTER 24 H, G	1,1273	1,1341	1,1900
	AFTER 48 H, G	1,1171	1,1260	1,1848
	AFTER 72 H, G	1,1161	1,1276	1,1864

TABLE 6.1 *(Continued)*

DAC (0,4)-CH	BEFORE TREATMENT, G	1,2004	1,1505	1,1154
	AFTER IMPREGNATION, G	3,8602	4,0128	4,4905
	AFTER 24 H, G	1,2067	1,1800	1,3346
	AFTER 48 H, G	1,1849	1,1676	1,3262
	AFTER 72 H, G	1,1989	1,1729	1,3327

*The number of aldehyde groups (mM/g carrier) is indicated in parentheses.

As seen from the obtained data in the absence of Gl the weight of matrix was recovered within 24 h, that is, the solvent is completely removed from the matrix. In the presence of 2 percent Gl, the solvent was not completely evaporated from the surface of a textile material after 72 h that will assist to the preservation of atraumatic properties of applications.

Thus, the addition of Gl to a wetting solution for modified cellulose dressings may be a promising method to maintain a moist environment. Further it was necessary to check the effect of Gl on the conservation of the activity of PC enzymes, which are incoming to the composition of modified DM. For this purpose we studied the changing of the PC activity (by azocoll) in dependence on the concentration of Gl in PB saline (Figure 6.1).

Besides, it was studied the influence of the different concentration of Gl on the enzymatic activity of various forms of PC: in solution, stabilized with Ch-gel, or immobilized on various cellulosic carriers containing Ch. The data obtained are also shown in Figure 6.1.

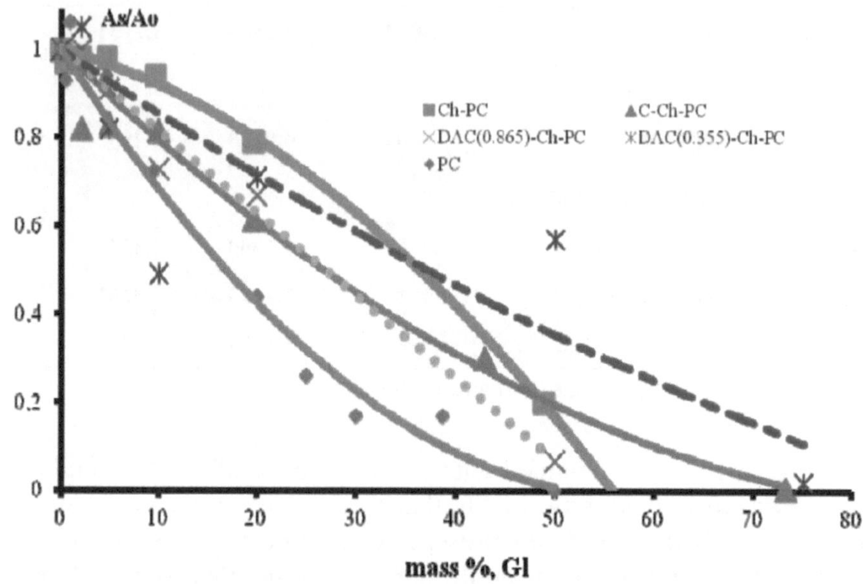

FIGURE 6.1 The dependence of the enzymatic activity of various forms of the PC (substrate: azocoll) on the concentration of glycerol. $A_s/A_0 = 1$ without Gl. As can be seen from the data of Figure 6.1, the use of recommended by us 2–5 percent of Gl solution reduces the activity of PC enzymes not more than 15 percent.

Hydration of proteins is one of the main factors affecting the folding of proteins and the expression of their functional activity. Low molecular weight organic compounds (including Gl) can act as regulators of enzyme activity, although it is possible as a direct impact of organic compounds on protein via intermolecular contacts and their indirect action through changes in the structure and dynamics of the hydration shell. Apparently, this explains the nonlinear change in enzyme activity in dependence on the Gl concentration (at low Gl concentrations (1–20%)). Understanding of the mechanisms of proteins' interaction with water and the third component is a fundamental task. In the applied aspect the number of organic solvents (including glycerol) allow to optimize biotechnological processes, significantly increasing the stability of enzymes, providing a reactionary contact to chemical compounds of different polarity, and so forth, that is often unattainable within a purely aqueous environment [21].

As is well known [11, 21], protein denaturation in aqueous–organic mixtures is caused by the destruction of the hydration shell of the protein molecule owing to competitive displacement of its constituent molecules of water per molecule of organic solvent. Consequently, we can assume that the increase of protein stability against denaturing effects of the organic solvent can be achieved by increasing of the strength of the hydration shell of protein to its surface. In turn, the ability of the surface to hold the hydrate shell is the higher the more hydrophilic it is. Thus, hydrophilization of the protein surface should lead to an increase of its stability to denaturizing in mixtures of water–organic mixtures [11]. In our case, the destruction of the hydration shell may be prevented by immobilization of the polysaccharide compound on cellulose carrier.

Previously, we have shown that after immobilization of PC on C or DAC (with different degrees of modification) [3, 17], there is a complete loss of enzymatic activity of the immobilized preparations within 1 year of storage. One of the reason of inactivation of protein is that low molecular aldehyde (e.g., formaldehyde) inactivates proteolytic enzymes.

To increase the shelf life of products synthesized by us, it was decided to use a carrier that has a positive charge; for this purpose, cellulose matrix (unmodified or periodate oxidized) was at first treated with a solution of chitosan, and then PC was immobilized on it [3, 17]. The amino group of chitosan readily reacts with aldehydes, halides, and other compounds having an active group [3, 5]. Figure 6.2 shows the scheme of obtaining of our preparations.

A – Immobilization of PC on DAC

DAC *PC* *DAC-PC*

B - Interaction of PC and Ch

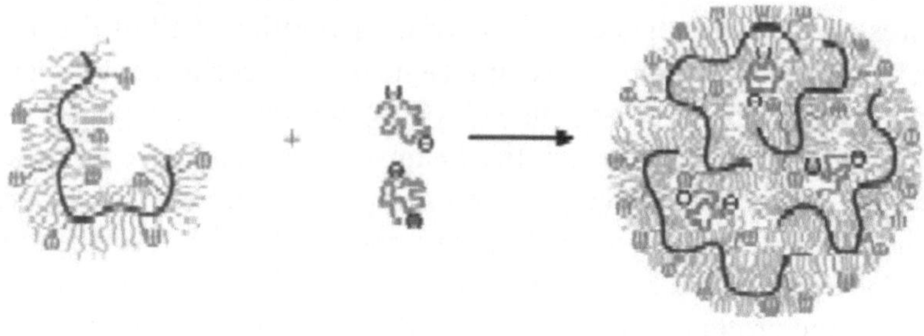

Chitosan PC Ch-PC

C – Immobilization of PC on DAC(C)-Ch

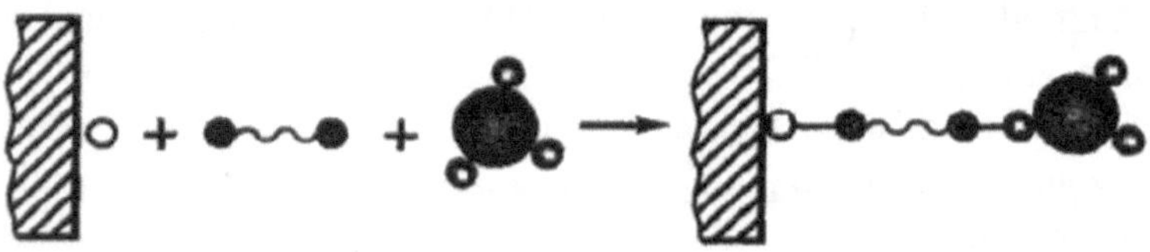

C or DAC Chitosan PC C(DAC)-Ch-PC

FIGURE 6.2 Schemes for producing of modified PC.

In was studied the changing of enzyme activity of obtained various PC-containing preparations (different forms of cellulosic matrix modified by Ch) in dependence on the concentration of Gl (Figure 6.1). From these data were determined value A_{50} glycerol concentration (wt %), in which 50 percent of enzyme activity (substrate azocoll) is retained (Table 6.2).

TABLE 6.2 A_{50} dependence on the used preparation

S. No.	Name of the Preparation	A_{50}; mass %
1.	PC	17.5
2.	Ch-PC	35
3.	C-Ch-PC	31.5
4.	DAC(0.355)-Ch-PC	65
5.	DAC(0.865)-Ch-PC	31.5

As was shown in [22], Gl molecules interact with Ch to form various adducts. Moreover, by increasing the concentration of Gl various "clusters" occur. Amino groups of Ch react irreversibly with C and DAC proportionally as carboxyl or carbonyl (aldehyde) groups on the fibrous carrier [9]. The remaining amino groups of Ch will be fully engaged "sour" groups (isoelectric point of PC enzymes is below 3) by PC proteins [25, 26]. Thus, in our view, free amino group on Ch absent in the derivatives of C (DAC)–PC–Ch. As evidenced by the absence of the shift of pH-optimum of enzymatic activity for the immobilized PC. The data obtained are shown in Figure 6.3.

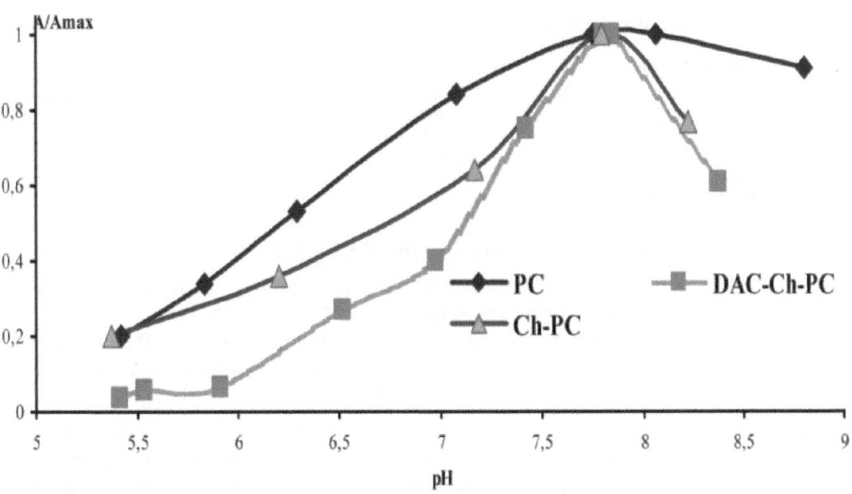

FIGURE 6.3 The pH dependence of the enzymatic activity of various forms of PC (substrate azocoll)

Stabilization of PC when exposed to Gl solutions is explained as immobilization of proteins on the carrier and interaction of the Ch molecules with Gl [22–24].

CONCLUSION

It is shown that impregnation even by 2 percent Gl-solution of cellulose applications does not allow completely to evaporate the solution (at 37°C for 72 h), which will facilitate atraumatic properties of used products.

Various forms of Ch stabilize enzymes belonging to the PC, protecting them from the denaturing action of glycerol. It is planned to investigate the effect of temperature and time to various forms of PC. Also it is noted that in addition to the proteinases there may be present enzymes which depolymerize Ch in the PC [22, 23].

KEYWORDS

- **Chitosan**
- **Chitosan-coated dialdehyde cellulose**
- **Glycerol**
- **Medical applications**
- **Modification**

REFERENCES

1. Mashkovskyi, M. D.; Medicinal Agent. 14th edition, **2002**, p. 250.
2. Tolstyh, P. I.; Gostishev, V. K.; and Arutunjan B. N.; et al. The proteolytic enzymes immobilized on fibrous materials in surgery/Erevan, Aystan, **1990**, p. 136.
3. Belov, A. A.; Textile materials containing immobilized hydrolases for medical and cosmetic purposes. Production. Properties. Application. Germany: LAP LAMBERT Acadmic Publication, GmbH & Co. KG; **2012**, 242 p.
4. Lukanina, K. I.; Development of scientific and technological bases of creation of dressing materials from the biodestruktive and biocompatible fibrous materials on a basis polylactide the Abstract theses on competition of a scientific degree of Candidate of Technical Sciences, Moscow, **2011**, 24 p.
5. Muzzarelli, R. A. A.; Chitins and chitosans for the repair of wounded skin, nerve, cartilage and bone. *Carbohyd. Pol.* **2009**, 76, 67–182.

6. Moskvichev, B. V.; and Poliyk, M. C; The immobilized enzymes. Serpukhov, M. **1992**, 112 p.
7. Wolf, M.; and Ransberger, K.; Treatment by Enzymes. Ed. Mir, M., **1976**, 46–47.
8. Nazarenko, G. I.; Sugurova, I.; Glyantsev, Yu.; Wound, S. P.; Bandage, patient. *M. Med.* **2002,** 469 p.
9. Simona Strnad, Olivera Šauperl and Lidija Fras-Zemlji.; Chapter 9/Cellulose fibres functionalised by chitosan: characterization and application. In: Biopolymers. Ed. El-nasha, M. **2010**, 181–200.
10. Belov, A. A.; Influence of water solutions of a dimethyl sulfoxide on not modified and immobilized enzymes. *Chem. Technol.* **2004,** *12,* 35–40.
11. Belova, A. B., Mozhaev, V. V.; and Levashov, A. V.; etc. Interrelation of physical and chemical characteristics of organic solvents with their denaturant ability in relation to proteins. *Biochemistry*. **1991**, *56*(11), 1923–1945.
12. Klyachko, N. L.; Bogdanova, N. G.; Martinek, K.; and Levashov, A. V.; Replacement of water by water and organic mix in systems of the turned micelles – a way to increase of efficiency of a enzymatic catalysis. *Bioorg. Chem.* **1990**. *5*(16), 581–589.
13. Segura-Seniseros, E. P.; Ilyin, A.; and Montalvo-Arredondo, H. I.; etc. Estimation of influence of pectin-papain interactions on the stability of the enzyme, and mechanical properties of pectin films, passion fruit, used to treat skin wounds. Vestn. Mosk. Un-ta. Ser. 2. *Chemistry*. **2006**, *47*(1), 66–72.
14. Martinek, K., and Semenov, A. I.; Kataliz enzymes in organic synthesis. *Achiev. Chem.* **1981**, 45(8), 1376–1406.
15. Rogovin Z. A.; Chemistry of cellulose. M. *Chemistry*, **1972**, 125–244.
16. Rutherford H. A., Minor F. W., Martin A. R. and Harris M.; Oxidation of cellulose: the reaction of cellulose with periodic acid. *J. Res. of the N. B. S.* **1942**. *29*, 131–143.
17. Belov A. A.; Filatov V. N.; and Belova E. N.; The medical bandage containing a complex of enzymes from hepatopancreas of a crab, and way of its receiving. Patent RF №2323748 from 21.02.06.
18. Belov, A. A., Ryltcev, V. V., and Ignatuk, T. E.; Methods of determining proteolytic activity in commercial samples of immobilized proteinase preparations. *Chem. Pharm. Jorn.* **1992**. *11–12*. 101–103.
19. Gaida, A. V.; Monastic, V. A.; Magerovsky, Yu. V.; and Danysh, T. V.; Sposob of determination of fibrinolitichesky activity. Patent of USSR 1255641 A1 USSR, (Opening. -1986. – N 37).
20. Kunitz, M.; Cristalline soubean trypsin inhibitor. *J. Gen. Physiol*, **1947**, *30*(1), 291–310.
21. Makshakova, O. N.; Modification of structure of a hydrate cover of polypeptides by aprotonny organic solvents: research by methods of IR-spectroscopy and quantum and chemical calculations. the thesis Abstract on competition of a scientific degree of Candidate of Biology, Kazan, **2010**, 24 p.
22. Quijada-Garrido, I.; Iglesias-Gonzarlez, V.; Mazorn-Arechederra, J. M.; Barrales-Rienda J. M.; The role played by the interactions of small molecules with chitosan and their transition temperatures. Glass-forming liquids: 1,2,3-Propantriol (glycerol). *Carbohyd. Pol.* **2007**, *68*, 173–186.

23. Duško Cakara; Lidija Fras; Matej Bracic; Karin Stana-Kleinschek; Protonation behavior of cotton fabric with irreversibly adsorbed chitosan: A potentiometric titration study. *Carbohyd. Pol.* **2009**, *78*, 36–40.
24. Lidija Fras Zemljič1, Simona Strnad, Olivera Šauperl and Karin Stana-Kleinschek; Characterization of Amino Groups for Cotton Fibers Coated with Chitosan. *Text. Res. J.* **2009,** *79*(3), 219–226.
25. Mukhin V. V., Novikov V.; Yu.; Rysakova, K. S.; Properties hitinoliticheskikh of enzymes hepatopancreas crab of Paralithodes Camtschatica. *Prikl. bikhy. Mikrobiol.* **2007**. *43*(2), 178–183.
26. Mukhin, V. V.; Novikov, V.; and Yu; Proteoliz and proteolititchesky enzymes in tissues of sea invertebrates animals. *Murmansk: PINRO*. **2002**, 118.
The work was supported by the Ministry of education and science of the Russian Federation, the state contract of 16.552.11.7046

CHAPTER 7

USING OF THE EPR SPIN LABELING FOR THE INVESTIGATION OF THE SYNAPTOSOMAL MEMBRANE FLUIDITY CHANGES UNDER DIMEBON INJECTION IN VIVO

N. YU. GERASIMOV, O. V. NEVROVA, V. V. KASPAROV,
A. L. KOVARSKIJ, A. N. GOLOSHCHAPOV, and E. B. BURLAKOVA

CONTENTS

7.1 AIM AND BACKGROUND

Membrane structure plays important role in the development of dementia [6]. Therefore, it was important to investigate the membrane structural changes under neuroprotector injection.

7.2 INTRODUCTION

In the last few years, antihistamine drug dimebon is proposed for the treatment of neurodegenerative disorders such as Alzheimer's disease. In [1], it was shown that the drug at low concentrations is AMPA-kainate receptors protogonist and NMDA-receptors antagonist, which action mechanism remains unknown. It is assumed that dimebon exhibits neuroprotective properties and improves cognitive function in dementia [2–4].

The lipids composition and lipid bilayer structure of membrane significantly affects on the proteins functional activity [5]. Earlier, we showed disorders in the membranes structural characteristics as a result of the Alzheimer's disease development [6]. In addition, Burlakova proposed membrane memory model [7], in which the determining factor is the structure of the membrane, and fluidity is one of the important structural characteristics of lipid bilayer. Therefore, it is important to study the neuroprotector's action on the structural state of the lipid bilayer; accordingly, this work investigated the Dimebon effect on the fluidity of mice synaptosomal membrane by EPR spin labeling method.

7.3 EXPERIMENTAL

Dimebon was kindly provided by Bachurin S. O., IPAC RAS (Figure 7.1). The drug was injected abdominally every day in the concentration 1 mg/kg. Samples were taken in 1, 3, 7, and 15 days after injection. Female of HLK white outbread mice 20–23 g in size were used as experimental animals.

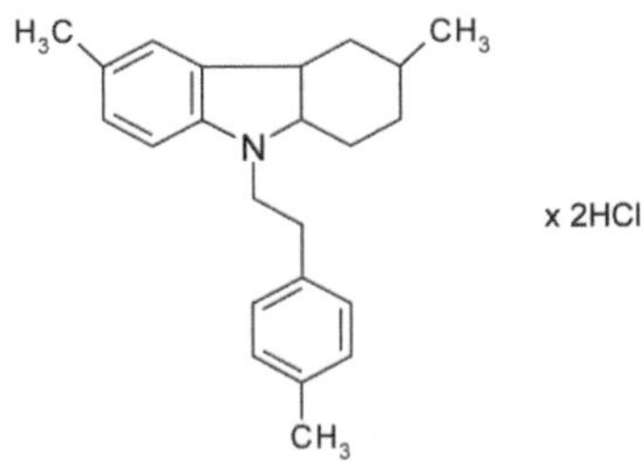

FIGURE 7.1 Dimebon.

The sample was a synaptosomes combined fraction isolated from the brain of six to eight animals. Each measurement was carried out 4–5 times. Synaptosomes were separated by differential centrifugation in sucrose [8].

The fluidity of the membranes lipid bilayer was determined by the method of electron paramagnetic resonance (EPR) of spin probes. As probes it were used the stable nitroxide radicals 2,2,6,6-tetramethyl-4-capryloyl-oxypiperidine-1-oxyl (probe I) and 5,6-benzo-2,2,6,6-tetramethyl-1,2,3,4-tetrahydro-γ-carboline-3-oxyl probe II), synthesized at the Institute of chemical physics named after N.N. Semenov RAS (Figure 7.2).

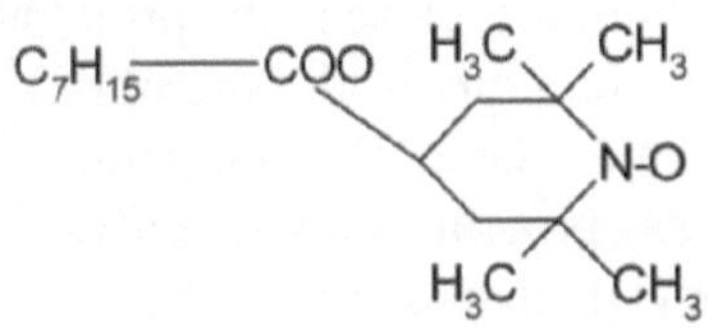

Probe I

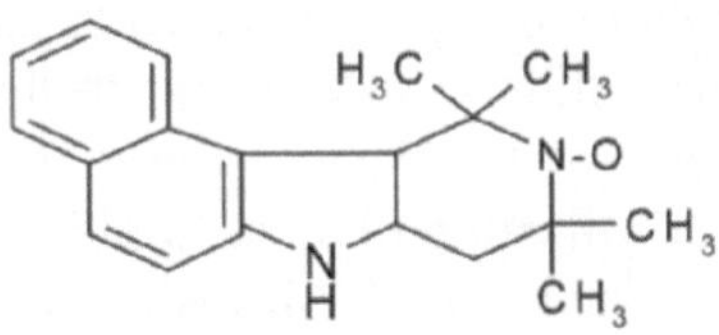

Probe II

FIGURE 7.2 Spin probes.

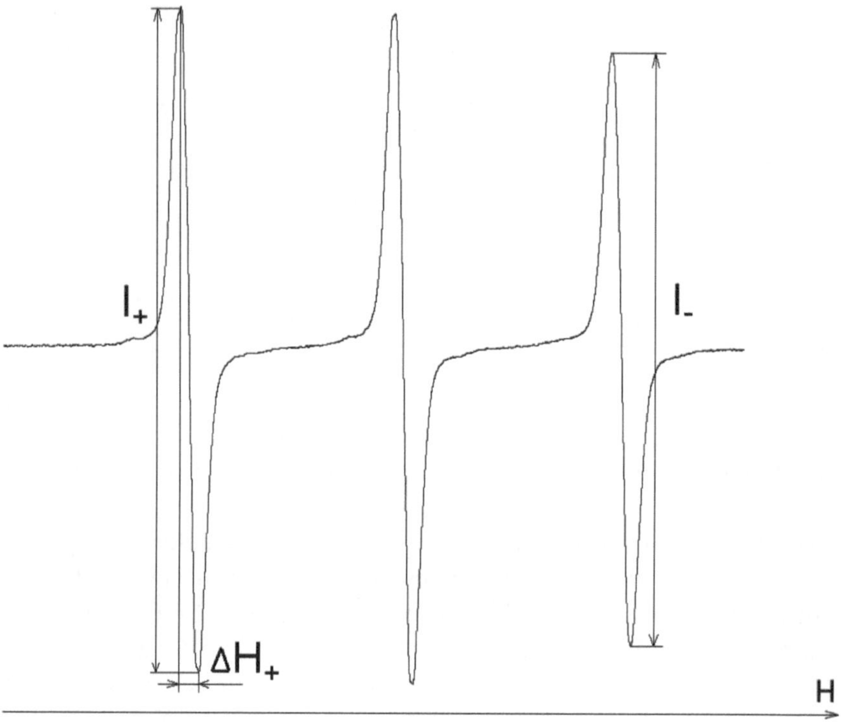

FIGURE 7.3 Typical ESR spectrum for probe I and II.

In [9], it is shown that the probe I is localized mainly in the surface layer of membrane lipid components, and probe II—in lipids, close to the proteins, that allow to talk about the lipid–protein interactions in the membranes based on the probes I and II behavior in the lipid bilayer. For convenience, we will call probe I as "lipid," and probe II as "protein" subsequently.

The rotational diffusion correlation time (τ_c), characterizing the membrane components microviscosity, was calculated from EPR spectra (Figure 7.3) by formula

$$\tau_c = 6.65 \times 10^{-10} \times \Delta H_+ \times ((I_+/I_-)^{0.5} - 1),$$

presented in [10]. The EPR spectra were registered in the temperature range of 283–317 K (10-44 C) on radiospectrometer ER 200D-SRC "Brucker" in the magnetic spectroscopy center of the Emanuel Institute of Biochemical Physics RAS.

The well-known relation by Stokes–Einstein (see, e.g., [11]) binds the parameter τ_c with medium viscosity, surrounding the probe, by formula $\tau_c = \eta V/kT$, where V is the volume of the radical (it can be considered directly proportional to the molecular weight); η is the dynamic viscosity of the medium; k is the Boltzmann constant, and T is the absolute temperature. Dynamic viscosity η is related to temperature by the following empirical relation $\eta = A'e^{b/T}$ [12], when it follows $\ln\tau_c = A'' + b/T + \ln(1/T)$, where A′, A″, b are the constants. Investigated in our work, temperature range (from 283 to 317 K) is sufficient narrow insomuch, that component ln (1/T) changes is very low in comparison with the term b/T, so we can assume $\ln\tau_c = a + b/T$.

Thus, the experimental curves should be linearized in the coordinates $\ln\tau_c$ and 1/T. However, like such behavior is taken place only for simple, one-component system. Membrane structures are systems, characterizing by the presence of thermoinducible structural transitions. Accordingly, dependency plot of $\ln\tau_c$ from 1/T for such structures must be a polyline, which break points are the points of structural transitions [13]. The slope coefficient of these lines allows to determine the transition activation energy $\Delta E_a = bR$ [14], where b is the slope coefficient of the corresponding straight-line section, and R is the absolute gas constant. Activation energy corresponds to the transition energy of 1 mole of membrane lipids [14].

Statistical data processing was carried out by the methods of parametric statistics using software Microsoft® Office Excel and Origin® 6.1 with the statistical reliability 95 percent.

7.4 RESULT AND DISCUSSION

In this paper, we studied the effect of neuroprotector dimebon on synaptosomal membrane microviscosity isolated from the mice brain in 1, 3, 7, and 15 days after chronic injection of drug. An example of the rotational diffusion correlation time dependence on temperature is shown in Figures 7.4 and 7.5, in coordinates ln (τ_c) and 1/T. As it seen from figures, the dependence is a polyline, consisting of inclined and practically horizontal sections. Horizontal sections correspond to generalized structural changes of corresponding membranes areas (near protein—for probe II, Figure 7.5, and lipid—for probe I, Figure 7.4). Two transitions are observed in both areas of membranes at temperatures 289–293 K (16-20°C) and 311–317

K (38–44°C) for the control group. The first transition is associated with alterations in the lipid phase, and the second—with the changes of the proteins structure [15, 16]. On the other hand, after chronic injection of Dimebon an additional structural transition is appeared in synaptosomal membranes near protein areas in the temperatures interval 297–301 K (24–28°C). This fact apparently is associated with the integration of the present substance in the protein structures, including receptors and channels [2], that leads to changes of membrane proteins structure and their lipid environment.

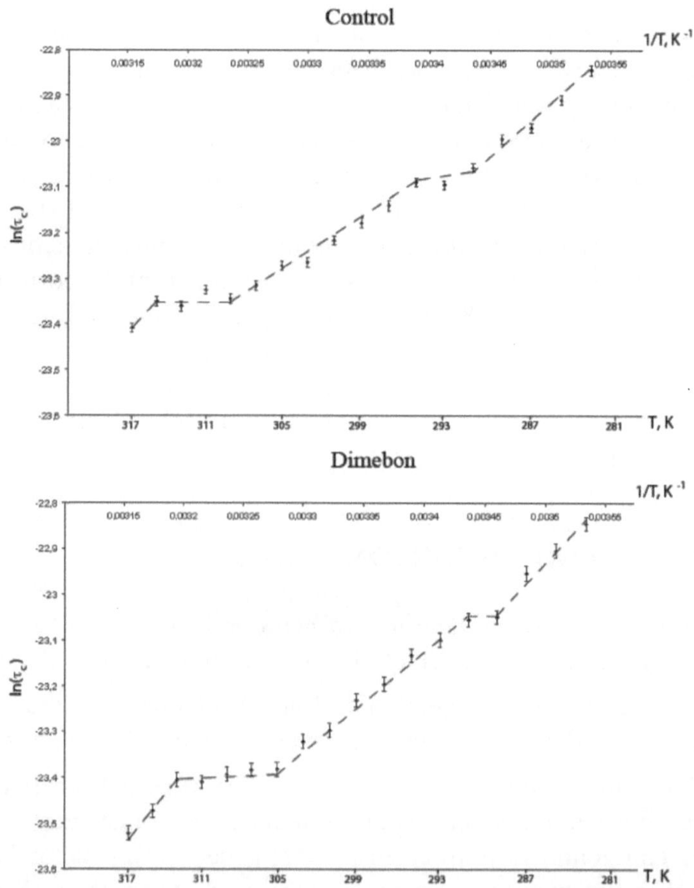

FIGURE 7.4 Dependence of the ln (τ) on 1/T of the probe I in synaptosomal membranes after 7 days of the Dimebon daily injections.

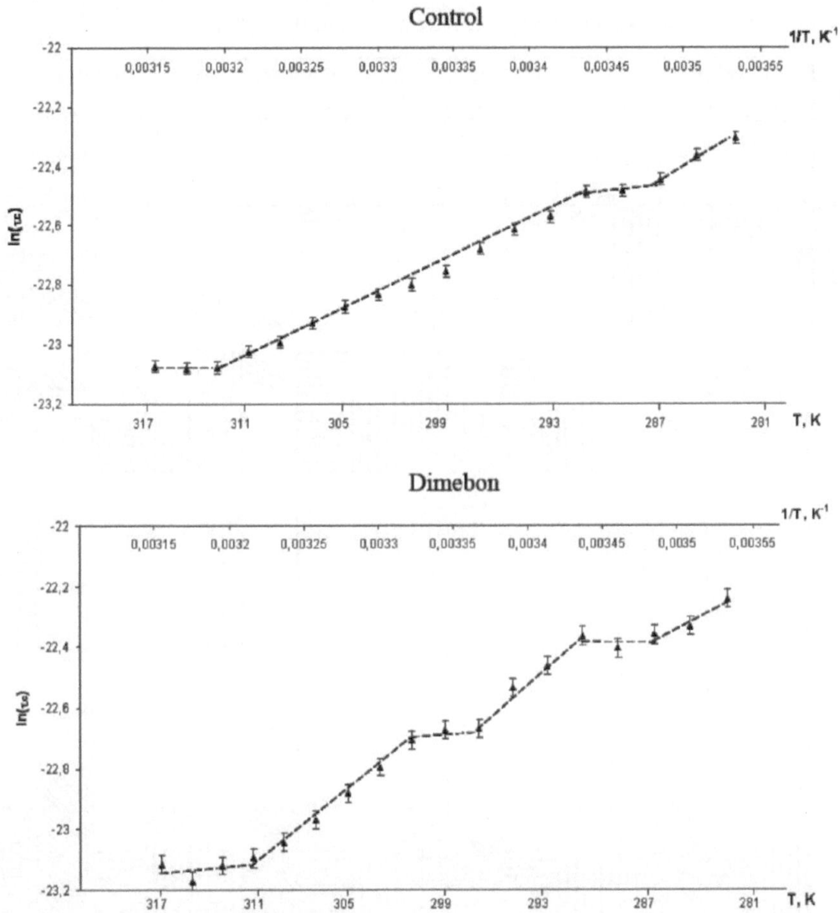

FIGURE 7.5 Dependence of the ln (τ) on 1/T of the probe II in synaptosomal membranes after 7 days of the Dimebon daily injections.

The lipid protein-free phase of membranes remained almost untouched (was not differ from the control). Possibly, Dimebon cannot get out easily from protein. Arising in this case strong bonds of dimebon with proteins leads to the appearance of additional structural transition in the membranes [2].

Table 7.1 shows the corresponding structural transitions for all periods of drug injunction with the activation energies.

TABLE 7.1 Structural transitions in the synaptosomal membranes under Dimebon daily injection with their activation energy. (*, statistical validity <90%)

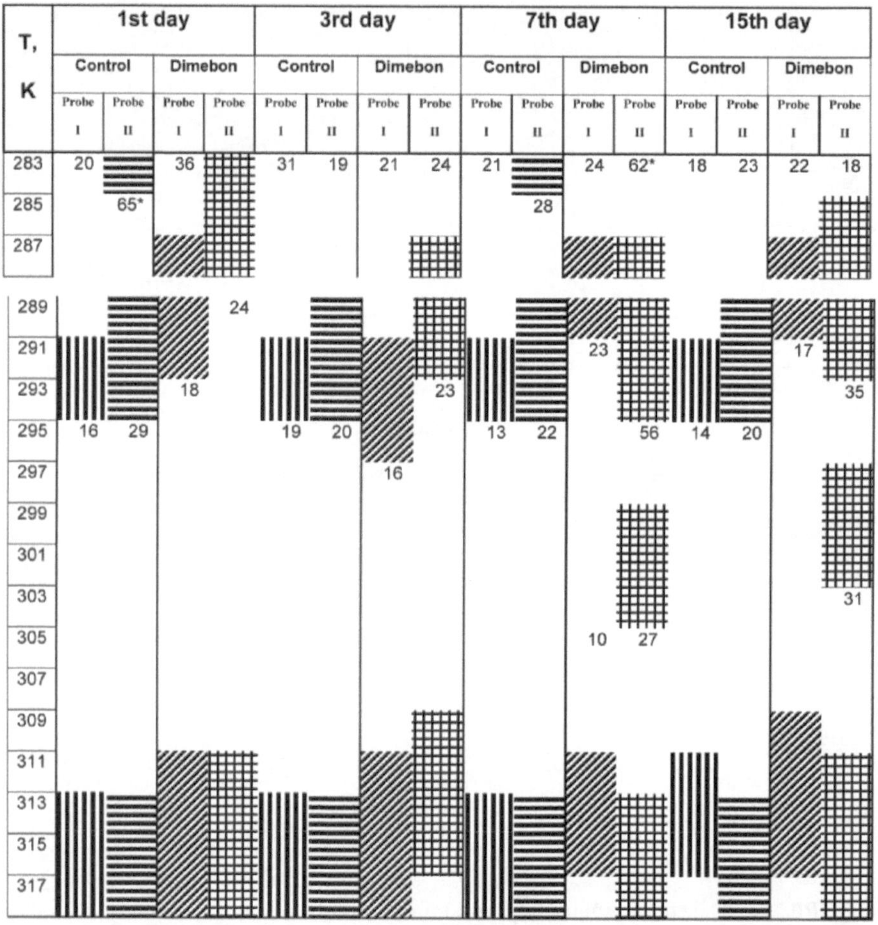

The table demonstrates that significant structural changes take place only at long-term delivering of agents compared with the control. For short-term drug injunction (1-3 days), the substance is excreted from the body and not accumulating in organism, without significant influence. Neuroprotector concentration is leveled off apparently at a constant value after the long periods of injection, whereby the effect from the drug action are increased. The activation energies of the corresponding transitions are not significantly changed and stay within the measurement accuracy close to the control, except the cases with additional transition. Additional transition appearance denotes a significant change of the membranes near protein areas structure that indicate about changes of the protein structure and their lipid environment.

The effect of the Dimebon on the synaptosomal membrane fluidity is shown in Figure 7.6. Values of the rotational diffusion correlation time at a temperature of 297 K, at which there is no thermoinduced structural changes in membranes, were taken as an indicator of microviscosity. The figure shows dynamic changes in near protein areas microviscosity with time, although these changes are minor in lipid areas.

The dependence of the relative changes of rotational diffusion correlation times from the agent introduction time at 297 K is shown on Figure 7.7. The fluidity of membrane near protein areas are distinctly increased immediately after the first injection of Dimebon that is a response to external action (Figures 7.6 and 7.7). Fluidity is returned to the level of control at the further injection.

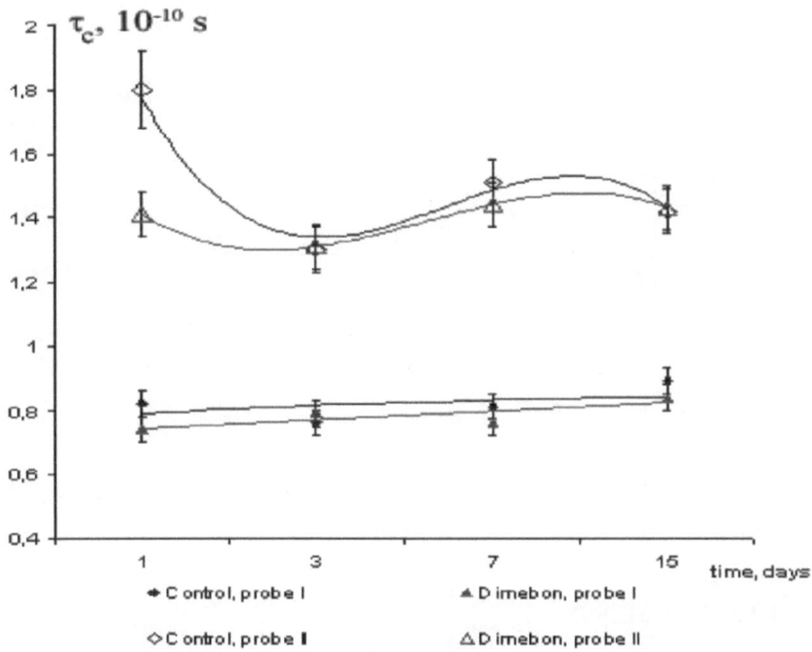

FIGURE 7.6 Dependence of the τ_c on time of the Dimebon daily injection for synaptosomal membranes at the temperature 297 K.

It is assumed that after a single injection organism is trying to exclude an foreign agent, and at the same time membranes structure is changed. An organism cannot removes the drug fully because of the strong bonds formation of Dimebon with proteins that leads to a change in the protein structure, and, as a consequence, lipid environment of protein changes its structure, in such a way that the structural characteristics are returned to norm. Subsequently, the body stops to responding on Dimebon injection, leading to an additional phase transition in the region 297–301 K (Figure 7.5, Table 7.1). In our opinion, such influence can lead to appearance of adverse effects with long-term use of the drug. Apparently, a certain level of fluidity is important for cells, because membranes lability was being returned to normal (Figures 7.6 and 7.7) with time.

Thus, we can assume that the membranes microviscosity is an important structural characteristics of membranes, and plays an important role in cell metabolism.

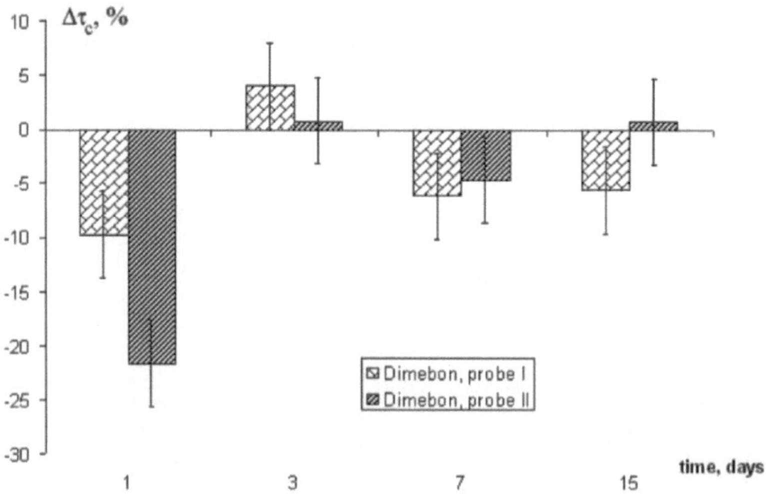

FIGURE 7.7 Dependence of the τ_c changes relative to control on time of the Dimebon daily injection for synaptosomal membranes at the temperature 297 K.

7.5 CONCLUSIONS

We have found that, neuroprotector Dimebon changes the membranes structure after chronic injection in such a way that the lipid bilayer fluidity is returned to normal with time. Therefore, membranes microviscosity plays an important role in the cell metabolism and is an important structural characteristic. Thus, at the drugs selection for the diseases therapy, the changes of the lability and membranes structure should be considered that probably will allow to avoid of adverse effects and to raise the effectiveness of medicinal product.

KEYWORDS

- **Dimebon**
- **Lipid–protein interactions**
- **Membranes fluidity**
- **Membranes structure**
- **Spin probe**

REFERENCES

1. Grigorev, V. V.; Dranyi, O. A.; and Bachurin, S. O.; Comparative study of action mechanisms of dimebon and memantine on AMPA- and NMDA-subtypes glutamate receptors in rat cerebral neurons. *Bull. Exp. Biol. Med.* November **2003,** *136(5),* 474–7.
2. Bachurin, S.; et al. Neuroprotective and cognition-enhancing properties of MK-801 flexible analogs. Structure-activity relationships. *Ann. NY Acad. Sci.* June **2001,** *939,* 219–36.
3. Grigoriev, V. V.; Proshin, A. N.; Kinzirskii, A. S.; and Bachurin, S. O.; Binary mechanism of action of cognition enhancer NT1505 on glutamate receptors. *Bull. Exp. Biol. Med.* July **2012,** *153(3),* 298-300.
4. Cano-Cuenca, N.; Solis-Garcia Del Pozo, J. E.; and Jordan, J.; Evidence for the efficacy of latrepirdine (dimebon) treatment for improvement of cognitive function: A meta-analysis. *J. Alzheimer's Disease.* **2013,** epub.
5. Robert, B.; Gennis: Biomembranes. Molecular Structure and Functions. New York: Springer-Verlag; **1997**.
6. Gerasimov, N. Yu.; Goloshhapov, A. N.; and Burlakova, E. B.; Structural state of erythrocyte membranes from human with Alzheimer's disease. *Khimicheskaja fizika.* **2009,** *28(7),* 82–86 pp. (in Russian)
7. Burlakova, E. B.; The role of the membrane lipids in the process of the information transfer. *Zhurn. fiz. khimii.* **1989,** 18(5), 1311 p., (in Russian)
8. Prohorova, M. I.; Metody biohimicheskih issledovanij. *Izd-vo Leningrad. un-ta.* **1982,** (in Russian).

9. Binjukov, V. I.; Borunova, S. F.; Gol'dfel'd, M. G.; i dr. Study of the structural transitions in the biological membranes using the method of spin probes. *Biokhimija.* **1971,** *36(6),* 1149 p. (in Russian)
10. Vasserman, A. M.; Buchachenko, A. L.; Kovarskij, A. L.; and Nejman, I. B.; Study of the molecular motion in the fluids and polymers using the method of the paramagnetic probe. *Vysokomolekuljarnye soedinenija.* **1968,** 10A, 1930 p., (in Russian)
11. Kuznecov, A. N.; Metod spinovogo zonda. M. Nauka, **1976,** (in Russian).
12. Kuhling, H.; Spravochnik po fizike. M.: Mir, **1983,** (in Russian).
13. Chapman, D.; Phase transitions and fluidity characteristics of lipids and cell membranes. *Quart. Rev. Biophys.* **1975,** *8,* 185–191.
14. Shinitzky, M.; and Inbar, M.; Microviscosity parameters and protein mobility in biological membranes. *Biochimica et Biophysica Acta.* **1976,** 133–149.
15. Gendel', L. Ja.; Gol'dfel'd, M. G.; Kol'tover, V. K.; Rozancev, Je. G.; and Suskina, V. I.; Investigation of the conformation transitions in the biological membranes using the method of the weakly bound spin probe. *Biofizika.* **1968,** *13(6),* 1114–1116pp. (in Russian)
16. Goloshhapov, A. N.; and Burlakova, E. B.; Thermo induced structural transitions in the membranes after antioxidants injection and with malignant grow. *Biofizika.* **1980,** *25(1),* 97–101 pp. (in Russian)

CHAPTER 8

A CASE STUDY ON THE INTERACTIONS MELAFEN AND IHFANS WITH SOLUBLE PROTEIN

O. M. ALEKSEEVA and YU. A. KIM

CONTENTS

8.1 INTRODUCTION

The main goal of this work was the investigation of the actions of two types of synthetics biological active substances to the soluble proteins that enriched the animal's blood serum. The first task was the test of the influence of plant growth regulator, applied in agriculture, Melafen, to the structural properties of soluble protein—bovine serum albumin (BSA). The second task was the investigation of the influence of hybrid antioxidants IHFANs that was expected to use as neuroprotectors, to the BSA structure properties

The first targets are the blood cells and the components of blood plasma also, when biological active substances appeared into the blood vascular system. This is why we carried out the serum albumin as the test for investigations of biological active substances actions. Serum albumins are water soluble globular proteins. Bovine serum albumin is a simple model of primary serum targets. The serum albumin has small size and, as the component that enriched blood plasma up to 50 percent, plays the essential role in the maintaining of osmotic balance sheets. As albumin enriches the blood sera, it facilitates the correct distribution of tissue liquid at many cases. The total area of all surfaces of albumins is biggest thanks to large amount of molecules and little molecular size of albumin. Besides this, the molecule may adsorb as hydrophilic, and the hydrophobe materials. These is why, the albumins are high effective carriers of most different molecules in blood plasma. And BSA takes essential part in transport of fatty acid, vitamins, hormones, and other materials that are needed to animal's organism good functioning.

So, the main purpose of our work was to determine how the aqueous solutions or emulsions of two types of synthetic biological active substances: Melafen and IHFANs, in a wide range of concentrations influence to the structure of soluble proteins with animals originated. Because of the present work, the action of Melafen and IHFANs on soluble proteins has been examined. By the obtained data, we may suppose or even predict how the biological active substances will be influenced on the protein's structure as well as protein–lipid interactions. Both these parameters are very important for a number of biological functions of albumins at animal body. So, if Melafen or IHFANs have some successful impacts on albumin molecular structure, which may be following by certain, changes of albumin properties, too. At this case, the transport function, or ability to support the osmotic balance will be changed, too.

Thus the selection of BSA as of experimental object was determined by the number of causes. Albumins have the famous structural and functional properties. The serum albumin is 50 percent from mass of all containing in blood sera of proteins. This is one of the first targets for biological active substances in blood serum composition. BSA is the perfect carrier for a numerous materials: Endogenous ones, like some free fatty acids, hormones, metal ions, bilirubin, and so forth, and exogenous ones (for example, materials that we want to test). Its structure is labile, and varies very easily. The molecular interaction of serum albumins with transported materials is determined of albumin's structural mobility, conditioned by the loop's stowage of one polypeptide chain, composed of 582 amino acid residues (Figure 8.1). Polypeptide chain forms nine loops that are fixed by 17 disulfide bonds. It is assumed that the polypeptide chain is laid in three more or less independent cooperative domains. One free SH-group exists in albumin molecule, which can take part in education of disulfides. Disulfides are at the core of trigger assembly of denaturation of this protein.

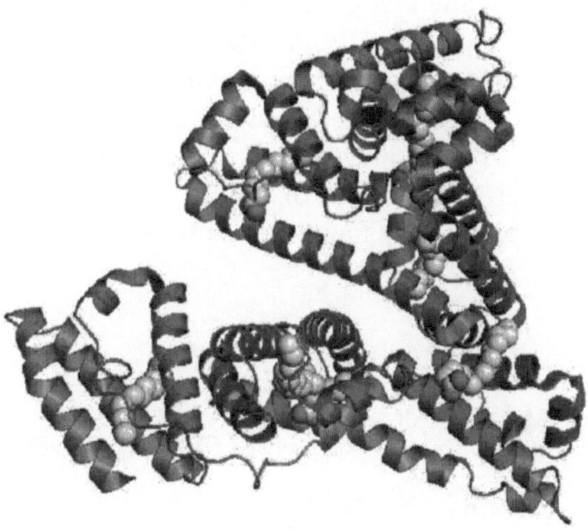

FIGURE 8.1 Scheme of serum albumin molecular structure.

Some changes of serum albumins conformation were registered on change of extent of quenching its intrinsic fluorescence. The numerous works is performed by this time, witch using of this approach for the test of actions of any biological active substances on albumins [1]. The albumin's binding with the exogenous synthetics materials we tested by using the registration of the intensity of intrinsic fluorescence of the BSA. Bovine serum albumin contains two tryptophane residues in hydrophobic regions of its molecule. There is the fluorescent emission of two tryptophane residues in hydrophobic regions of molecule BSA after excitation of tryptophane. First residue is located with close to a surface, second residue located at the deep inside of the protein globule. When the BSA molecule loosening, or unfolding, the availability of tryptophane residues for quencher-oxygen, which was dissolute in water, increase greatly. The quenching of tryptophane fluorescence was observed at this case. These changes of BSA tryptophane fluorescence intensity we registered with or without Melafen or IHFANs, when the wide concentration's region. And on the base of these data we may conclude, what Melafen- or IHFANs-aqua solutions or emulsion, under the certain concentration's region, can influence to the BSA structure. Than we may suppose how our tested material and under what region of concentration may have any influence to the functional properties of BSA.

The first task was the test of the influence Melafen to the structural properties of soluble proteins, BSA. Melafen is a plant growth regulator—heterocyclic organophosphor compound, synthesized at the A. E. Arbuzov Institute of Organic and Physical Chemistry of RAS. Melafen is the melamine salt bis (oximethyl) phosphinic acid. It was acquired by one stage with high stepping out of industrially available products [2]. Melafen is a hydrophilic poly functional substance (Figure 8.2).

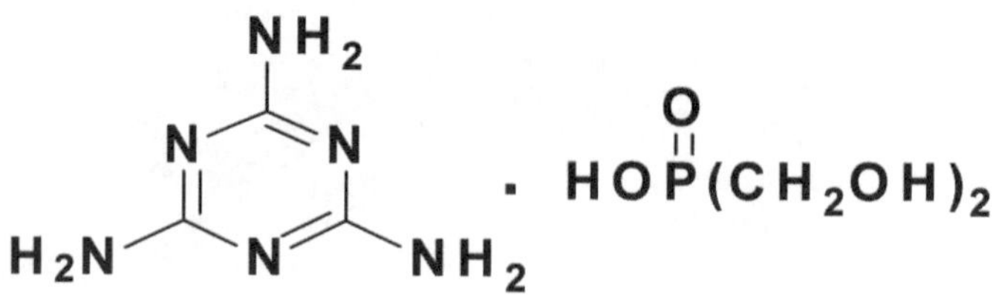

FIGURE 8.2 The structural formula of Melafen molecule.

Melafen raises the plants stress-resistance in the conditions of overcooling and drought, increasing the effectiveness of energy metabolism. In this case, Melafen causes the change of the fatty acid composition and the microviscosity of microsome and mitochondrial membranes in vegetable cells [4, 5]. Melafen is the strong regulator of plants stress tolerance under the bed environment. Aqueous solutions of Melafen at concentration 10^{-11}–10^{-9} M increased the plant growth, but the raising of concentration of Melafen up to 10^{-8}, 10^{-7} M leads to plant's seeds dye. Therefore, our studies were carried out in a wide range of concentrations (10^{-21}–10^{-3} M).

Considering the close interdependence of plant's and animal's bodies in nature, it was necessary to investigate the action of plant growth regulator at any objects of animal origin. The primary targets for biological active substances in animal's cells are membrane and their components. Performed analysis of actions of aqueous solutions of Melafen to the structural and functional characterizations of lipids and protein that built into the cellular membranes [6–8], was complemented by the testing of Melafen influences on free soluble proteins unbounded with membranes. As such model the protein of BSA was used.

The second task was the investigation of the influence of hybrid antioxidants IHFANs to the BSA structure properties. IHFANs were expected to use as neuroprotector. For the experiments, IHFANs were used as the aqua-ethanol suspensions at the wide concentration range (10^{-21}–10^{-3} M). IHFANs are the derivatives of antioxidant phenozan (β-(4-hydroxy-3, 5-di-tert-butylphenyl)propionic acid). Phenozan was created for stabilization of polymer at Institute of Chemical Physics of RAS Moscow [9]. It is known that the antioxidants often shall be used for the therapy of any pathological states. The derivative of phenozan—its potassium salt, was tested as biological active materials. It was turned out that potassium salt of phenozan exhibited the property of strong antioxidant and structural effectors on enzymes and on biomembranes [10]. However, the phenozan did not have the certain targets for its actions at membrane. As, the amphiphilic agent, phenozan acts primarily at all regions of surface layers of biomembrane, as in exterior and in internal sheets of bilayer. It was appeared, that phenozan penetrated through biomembrane defects to internal surface of bilayer plasma membrane, that was discovered with using of spin-labeled EPR probes on erythrocyte membranes [11]. For ingress into deeper layers of membrane without defects, it was necessary synthesized the more hydrophobic antioxidant. For that purpose, the choline esters was added to phenozan that was quaternized by long chain alkyl halogenides with number of carbon atoms from 8 up to 16-(4-hydroxy-3, 5-di-*tert*-butyl phenyl)propionyl butyl] ammonia halogenides that were named IHFANs

[12]. The series of hybrid multitarget antioxidants—IHFANs for orientation of antioxidant action, had been synthesized in IBHF RAS. Constructions were biological active. Its was based on phenozan with conservation of screened phenol. And one choline residue and one alkyl residues of different lengths (C8–C16) were inserted in that the complex molecules. So that molecules received antioxidant activity, and bought the new activities: the anticholinesterase activity, and IHFANs received the ability to penetrate bilayer by introducing of alkyl residues of different length (C8–C16) into the hydrophobic regions [13]. Scheme of IHFANs is presented at Figure 8.3.

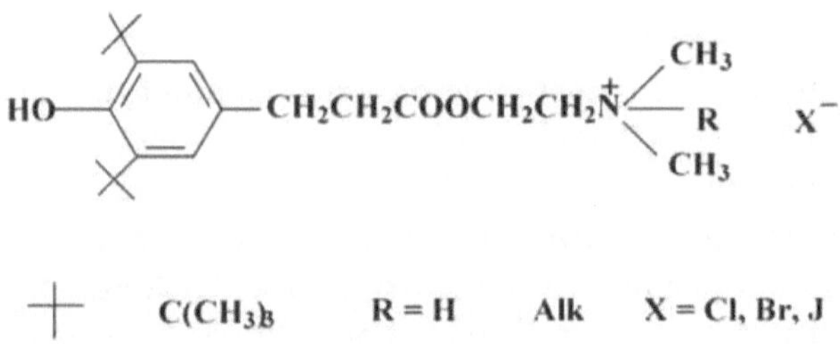

FIGURE 8.3 The structural formulas of IHFANs molecules. (C8) R =C_8H_{17}; (C10) R= $C_{10}H_{21}$; (C12) R= $C_{12}H_{25}$; (C16) R= $C_{16}H_{33}$; X = Br-.

As was shown at Figure 8.3, the hybrid antioxidants—IHFANs, have a charged onium group and a lipophilic long-chain alkyl tail. These structures of complex molecules allowed them to interact effectively with a charged lipid bilayer, insert to hydrophobic regions of cell membranes and maintain the antioxidant status. These molecules are bounded on membrane surface by the positive charge on quaternary nitrogen (the anchor), and the alkyl residue (the float) is introduced in bilayer being disposed among fatty acid's residues of phospholipids. And alkyl halogenides with variable length that were added to phenozan: R-C8H17; C10H21; C12H23; C16H33, were implemented to the membrane bilayer on the different deeps. Thus, ICHFANs localized in anion heads regions and in fatty acids tails. This phenomenon fortifies the membrane structure so much. And the membrane became resistant to any bad environment actions.

Performed analysis of actions of aqueous suspension of IHFAN-C-10 to the structural and functional characterizations of protein, that built into the cellular membranes—the erythrocytes and its ghost [14, 15], were complemented by the influence testing of ICHFANs on free soluble proteins unbounded with membranes. As such model was used the protein of BSA. These experiments of IHFANs influences to BSA structure, were similar as experiments of Melafen testing were performed with aid of quenching if intrinsic fluorescence method.

8.2 MATERIALS AND METHODS

The materials: BSA (Sigma). Melafen [melamine salt bis(oximethyl)phosphinic acid] was synthesized at the A. E. Arbuzov Institute of Organic and Physical Chemistry of RAS Kazan. IHFANs (4-hydroxy-3, 5-di-*tert*-butyl phenyl) propionyl butyl] ammonia halogenides were synthesized at the Institute of Biochemical Physics of RAS Moscow.

Bovine serum albumin was used as the aqua solutions. Melafen was used as the aqua solutions at the wide concentration range (10^{-21}–10^{-3} M). IHFANs were used as the aqua-ethanol suspensions at the wide concentration range (10^{-21}–10^{-3} M).

The standard methods with the standard conditions had been used for measurements of fluorescence quenching: 1 mkM of BSA protein aqua solution, 20°C. The quartz cell (1 sm.) was used for BSA (with or without Melafen or IHFANs) fluorescent intensity registrations by fluorescent spectrophotometer "Perkin-Elmer MPF-44B." The spectrophotometer "Specord M 40" was used for measurements of optical density of BSA protein aqua solutions when low concentrations.

8.3 RESULTS AND DISCUSSION

The first task of this work was the supporting of the Melafen influence to the albumin structure. Melafen at this case was the one of the factor of the certain variable bed environments. It is clear that the probability of albumin molecules to release and absorb the fatty acids and another absorbed substances was under the strong influence of environment. Melafen increase the crop-producing power of vegetables and seeds. The plant cells drastically increased of metabolism upon the treatment of small doses of Melafen that followed to greet elevating of plant's resistance to difficult environment. For resolve of the Melafen–BSA interrelation, we provided the spectral analysis. Data of the dependence of optical density from varied concentrations of Melafen are shown in Figure 8.4.

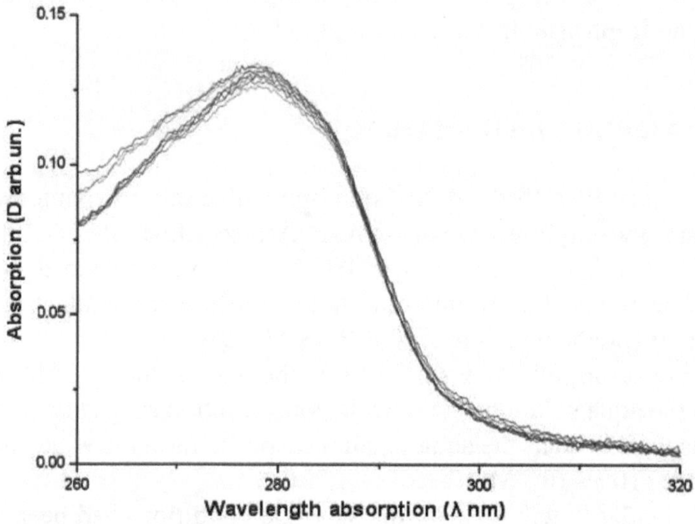

FIGURE 8.4 The spectrums of BSA optical density. The dependence of optical density from varied concentrations of Melafen.

As shown in Figure 8.4, the effect of aqueous solutions with varied concentrations of Melafen on spectrum patterns of BSA and the changes in absorption spectrums of BSA were negligible. The shape of BSA absorption spectrums, the locations of maximums of absorption didn't change drastically, when concentrations of Melafen were varied. Maximum of absorption spectrum BSA did not not shift; however, occurred the some change of the absorption degree and of spectrum shapes. Data of the dependence of correlation of BSA optical density with Melafen to without Melafen (D_{mel}/D) from Melafen concentrations were shown in Figure 8.5. The correlation (D_{mel}/D) was under the polymodal small changing, when Melafen concentrations varied.

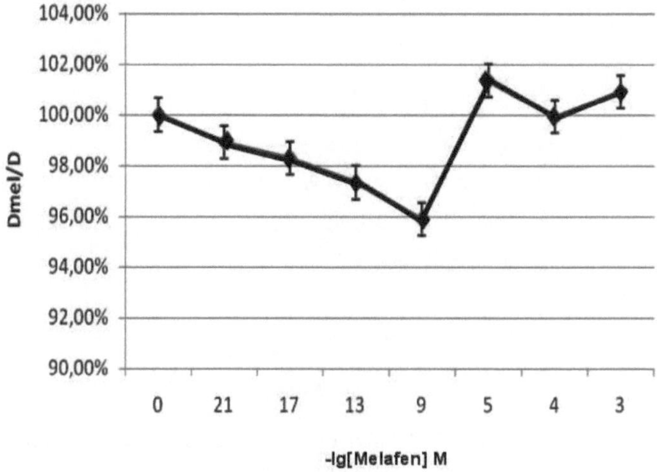

FIGURE 8.5 The dependence of correlation of BSA optical density with Melafen to without Melafen from Melafen concentrations.

Data of Melafen influence to the BSA absorption spectrums, which are shown in Figures 8.4 and 8.5, provides the evidence about absence of the covalent linkage between molecule Melafen and protein BSA. However, in registration of fluorescence spectrum, shown in Figure 8.6, revealed facts, indicative of great influence of aqueous solution of Melafen over a wide range of concentrations on conformation of BSA molecule. All solutions under the different concentrations of Melafen were not displaced the wavelength of fluorescence maximum. However, the fluorescence intensity has undergone a change. The great quenching of tryptophan fluorescence, when 10^{-4} M of Melafen was found, and the burst of fluorescence intensity when 10^{-17}–10^{-10} M was found too (Figures 8.6 and 8.7). The changes of fluorescence tryptophan intensity residues BSA are shown in Figure 8.6.

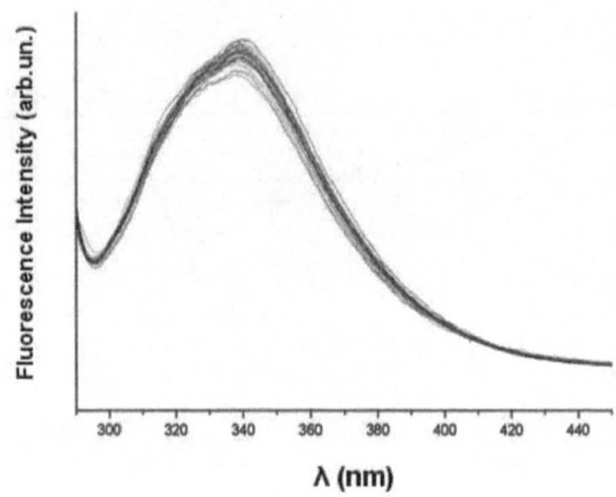

FIGURE 8.6 The influence of Melafen over a wide range of concentrations to the intensity of emission spectra of BSA tryptophan residues.

The data that fluorescence of BSA tryptophane residues was quenched by Melafen under the wide region of concentration were shown in Figure 8.7.

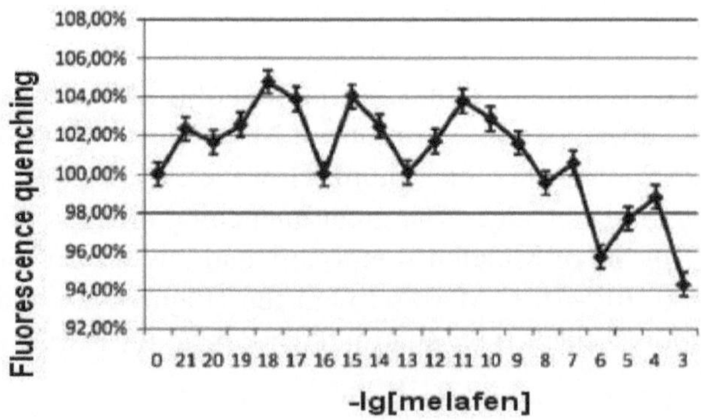

FIGURE 8.7 The Melafen influence to the fluorescence intensity of BSA. The control samples in absentia of Melafen have been adopted the fluorescence intensity as 100 percent. More 100 percent—is some burst of fluorescence intensity, less—is the quenching of fluorescence intensity.

As it can be seen from comparison of data, shown in Fig. 6, the shapes of emission spectra were similar, only the maximal value of fluorescence intensity was change in dependence from the Melafen concentrations. Then, we build a curve of the dependence of value at the maximum of tryptophan emission from Melafen concentration shown in Figure 8.7. We obtained the noticeable tendency of the fluorescence quenching by the Melafen, when large concentrations. And the some increasing of fluorescence intensity was occurred when low and ultrasmall concentrations of Melafen presented at the experimental medium. The –dose-dependence was polymodal, which is representative for biological active substances, effectual in small and ultrasmall doses [16]. Evidently, conformational rearrangements occurred in BSA molecules. These rearrangements were small and had the different direction.

The second task of our work was the investigation of the influence of hybrid antioxidants IHFANs that was expected to use as neuroprotector, to the BSA structure properties. IHFANs were used as the aqua-ethanol suspensions at the wide concentration range (10^{-21}–10^{-3} M). The spectral analysis of BSA–IHFANs interrelationships was occurred similar as for testing of Melafen–BSA actions.

The soluble protein the BSA in the presence of large concentrations of testable materials was turning, and becomes greatly accessible to water introduction. In the presence of low concentrations, on the contrary, the protein structure appears to much be getting stronger. The alkyl-halogenides tails of IHFANs are adsorbed on protein and defend it from molecular untwisting and water ingestion. The degree of protection was depended on length of alkyl tail directly in proportion. The maximum of protection has been found for IHFAN-C16 relationships with BSA (Figure 8.8).

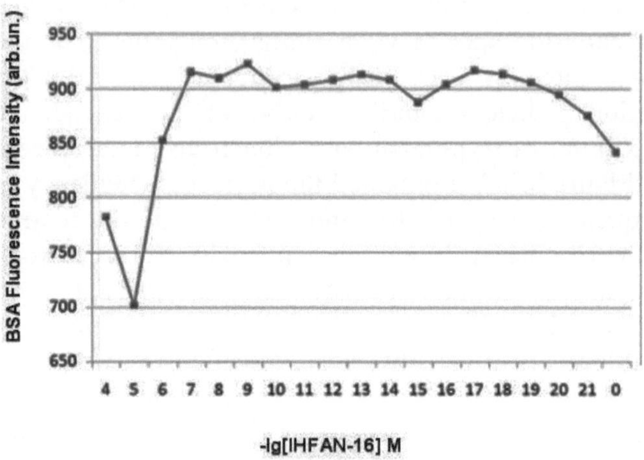

FIGURE 8.8 The influence of IHFAN-C16 to the tryptophan fluorescence intensity of BSA dependence of fluorescence intensity BSA from concentration IHFAN-C16 (10^{-21}–10^{-3} M).

The proteins, in which tryptophanes are contained, absorb the light near 280 nm, and the fluorescence spectrums are shifted to the short wavelength in contrast with tryptophane spectrums in water. Respectively importance their maximums may change from 343 nm (the serum albumin). Primary cause of this is the processes presence of orientation interaction, in which the spectrum regulation is determined by the polarity and microenvironment rigidity of chromophore. The dislocation of a high of the short wavelength is common to no polar environment, in long-wave for polar. The emitting maximums of proteins reflect average the availability their tryptophane residues in aqueous phase.

As was shown in Figure 8.9, the emitting maximums of BSA were shifted under the varied IHFAN-C16 concentrations. This shift was bigger when IHFAN concentrations were 10^{-6}–10^{-3} M. Are likely IHFANs "stick all over" the molecule of BCA, and thus the environment of tryptophane residues becomes more hydrophobe, then at IHFANs absences or when low their concentrations existence.

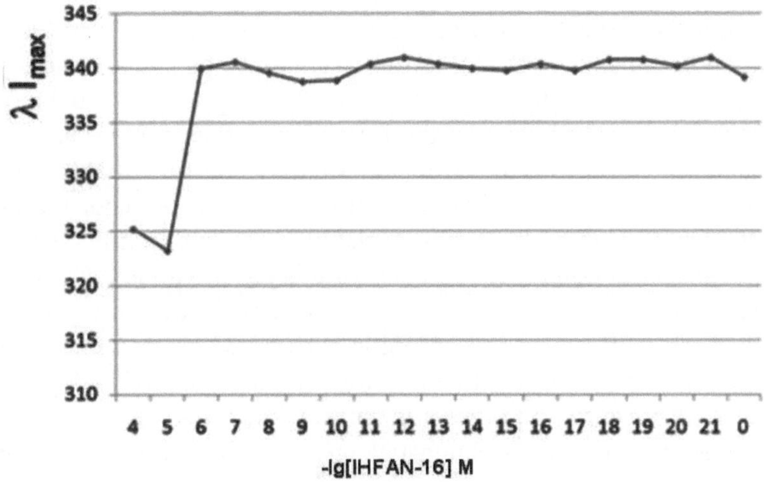

FIGURE 8.9 The influence of IHFAN-C16 (10^{-21}–10^{-3} M) to the tryptophane fluorescence intensity of BSA. The dependence of changes of wavelength of BSA fluorescence maximum from concentration IHFAN-C16.

When we test the IHFANs actions at the soluble protein BSA (Figures 8.6 and 8.7), we found the decreasing of tryptophane fluorescence when 10^{-4} M and the raising when 10^{-17}–10^{-7} M. For IHFAN-C16, which have the greatest alkyl halogenides tail, the inflammability had been greatest. Are likely, IHFANs communicate with albumin in low concentrations, adsorbed at BSA surface. So IHFANs when low and ultralow concentrations defend tryptophane residues from contacts with the water. And when large concentrations, IHFANs change the BSA structure, are likely, and increasing the availability of tryptophane for water. Some oxygen, which was solute at water, quenched the intrinsic tryptophane fluorescence of BSA molecules. But IHFANs, when the large concentrations (10^{-5}–10^{-3} M), not only decreased the intrinsic BSA fluorescence, but IHFANs shifted the fluorescence maximum to more short-wavelength. This indicates the initiation of more polar environment for BSA tryptophane residues.

8.4 CONCLUSION

Evidently, Melafen molecules affected to the albumin so that under the small and ultrasmall concentrations there was the preserving of the protein tryptophane residues from quenching from oxygen, dissolved in water. And under the large Melafen concentrations, the change of protein conformation became so essential. In this case, the tryptophane residues that lying at deep molecule locus became more available to water (and oxygen, respectively), on that indicated the fluorescence quenching. Occurs "the loosening" of BSA molecule structure. We may conclude that the soluble proteins that unhardened of the membrane lipids were under the essential Melafen actions. Considering Melafen is the hydrophilic substance, it can change the water environment. In this case, we may suppose that Melafen influence to BSA by two ways: mediated through the water, or directly to the influence to hydrophilic sites of BSA molecules. Mechanism was unknown. These influences were mainly changed in dependence on Melafen concentration present in surrounding solution. There were no clear evidences of BSA–Melafen linkage existence. However, mediated action through the change of water medium appears to occur surrounding the protein's molecules.

Also the water solutions of Melafen may be the regulator of transporting function of albumins, as its will be introduced to the animal's body. And it may be take part in extracting fatty acids from any molecules, or bounding of free fatty acids by albumins. As it is known, the water solutions of Melafen change the fatty acid's content of membranes [5].

The IHFANs actions on the BSA structure were different from Melafen actions. Evidently, because the Melafen is the simple neutral hydrophilic substances, but the IHFANs are the complex substances with hydrophobic part—long-chain alkyl halogenides, and hydrophilic part—positive charge on quaternary nitrogen. By this molecules of IHFANs are bounded on BSA surface by the positive charge and introduced to the deep locus of BSA by chains alkyl halogenides. As shown in our paper, the IHFANs preserve BSA molecule from "loosening" of molecule BSA structure. And the region of this nondestructive IHFANs concentrations was extended more (10^{-21}–10^{-7} M), than for Melafen nondestructive concentration's region (10^{-21}–10^{-9} M). The degree of protection was depended on length of alkyl tail directly in proportion. The maximum of preserving has been found for IHFAN-C16. It may be supposed that alkyl halogenides tails what having occupied all hydrophobic regions in albumin molecule, in practice came to form the containment shell from water encroachment to tryptophane residues. The oxygen, which was dissoluted in water, failed to penetrate to tryptophane residues. Thus the mechanisms of BSA–Melafen or BSA–IHFANs relationships were different. There were not the covalent binding, but the grade of absorption, points for absorptions were different. The "loosening" of molecule BSA structure occurred, when the large concentrations (10^{-5}–10^{-3} M) of Melafen were presented at medium for intrinsic BSA fluorescence registration. And IHFANs, when the large concentrations (10^{-5}–10^{-3} M), not only decreased the intrinsic BSA fluorescence. IHFANs shifted the fluorescence maximum to more short wavelength. This indicates the initiation by IHFANs of more polar environment for BSA tryptophane residues.

Obtained data in this work suggests to the fact that one of the first targets in blood, in particular, in blood plasma, was exposed to as Melafen, and IHFANs, used over a wide range of concentrations. The albumin structure was varied under the presence of Melafen or IHFANs. Respectively, the presupposition was arising that if some properties of albumin change, as carrier of biologically active substance and as osmotically active substance. Albumins maintain the colloid-osmotic pressure at blood plasma and at other fluids (e.g., in cerebrospinal fluid). It can be assumed that the transport effectiveness will be decreased and can be unbalanced of osmotic pressure in compartments with biological fluids. This is why the application these materials demands the great cares and the observances of concentration limitations, because the soluble protein of animal origin—BSA, changes his structural properties in their attendance so much.

KEYWORDS

- **Bovine serum albumin**
- **Fluorescence**
- **Hybrid antioxidant**
- **Melafen**

REFERENCES

1. Diaz, X.; Abuin, E.; and Lissi, E.; "Quenching of BSA intrinsic fluorescence by alkylpyridinium cations its relationship to surfactant-protein association". *J. Photochem. Photobiol.* **2003,** *155,* 157–162.
2. Fattachov, S. G.; Reznik, V. S.; and Konovalov, A. I.; "Melamine Salt of Bis (hydroxymethyl) phosphinic Acid. (Melaphene) As a New Generation Regulator of Plant growth regulator". In set of articles. Reports of 13th International conference on chemistry of phosphorus compounds. S. Petersburg; **2002,** S. 80.2.
3. Kostin, V. I.; Kostin, O. V.; and Isaichev, V. A.; "Research results concerning the application of Melafen when cropping". "Investigation State and Utilizing Prospect of Growth Regulator 'Melafen' in Agriculture and Biotechnology." Kazan; **2006,** 27–37.
4. Zhigacheva, I. V.; et al. "Influence of phosphoorganic plant growth regulator to the structural characteristics of membranes plant's and animal's origin". *Biol. Memb.* **2008,** *25(2),* 150–156.
5. Zhigacheva, I. V.; et al. "Fatty acid's content of mitochondrial membranes of Pea seedlings in conditions of insufficient moistening and treatment by the phosphoorganic plant growth regulator". *Biol. Memb.* **2010,** *27,* 256–261.
6. Alekseeva, O. M.; et al. "Melafen influence on structural and the functional state of liposomes membranes and cells of ascetic Ehrlich carcinoma" *Bull. Exp. Biol. Med.* **2009,** *147(6),* 684–688.
7. Alekseeva, O. M.; et al. "The Melafen-Lipid-Interrelationship Determination in phospholipid membranes". Doklady Akademii Nauk. **2009,** *427(6),* 218–220.
8. Alekseeva, O. M.; "The Influence of Melafen - Plant Growth Regulator, to Some Metabolic Pathways of Animal Cells". *Polym. Res. J.* USA, **2013,** *7(1),* Chapter 6, 15–23.
9. Ershchov, V. V.; Nikiforov, G. A.; and Volod'kin, A. A.; "Space screened phenols" M. Chemistry. **1972,** 352 p.
10. Burlakova, E. B.; Goloshchapov, A. N.; and Treschenkova, J. A.; "Action of low doses of phenosan on biochemical properties of lactate dehydrogenase and membranes microviscosity by the microsome of mouse brain". *Rad. Biol. Radioecology.* **2003,** *3,* 320–323.
11. Gendel, L. J.; Kim, L. V.; Luneva, O. G.; Fedin, V. A.; and Kruglakova, K. E.; "Changes of cursory architectonics of erythrocytes under the impact of synthetic antioxidant phenosan-1". *Izvestiya RAS Series. Biol.* **1996,** *4,* 508–512.
12. Nikiforov, G. A.; Belostockaya, I. S.; Vol'eva, V. B.; Komissarova, N. L.; and Gorbunov, D. B.; "Bioantioxidants 'Of float' type on the biologacal active substancesis of derivative 2, 6 ditertbutil fenol". Set of articles. "Bioantioxidants," Scientific Medical Academician, Tyumen. **2003,** 50–51 p.
13. Burlakova, E. B.; Molochkina, E. M.; and Nikiforov, G. A.; "Hybrid antioxidants". *Chem. Chem. Technol.* **2008,** *2(3),* 163–171.
14. Alekseeva, O. M.; Kim, Yu. A.; Rikov, V. A.; Goloshchapov, A. N.; and Mill, E. M.; "Influence of screened phenols on lipids structure, and also soluble and membrane proteins" in the book "Phenolic compounds: Fundamental and applied aspects," pub-

lishing house "Nauchnii Mir." (ISBN) Chapter 1, Phenolic compound: Structure, property, biological activity. **2010,** 116–126 p.
15. Albantova, A. A.; Binyukov, V. I.; Alekseeva, O. M.; and Mill, E. M.; The investigation influence of phenozan, ICHPHAN-10 on the erythrocytes *in vivo* by AFM method. In: "Modern Problems in Biochemical Physicsew Horizons". Varfolomeev, S. D.; Burlakova, E. B.; Popov, A. A.; Zaikov, G. E.; eds. New York: Nova Science Publishers; Chapter 5, **2012,** 45–48 p.
16. Burlakova, E. B.; "Effect of ultrasmall doses". *Vestnik RAS.* **1994,** *64(5),* 425–431.

CHAPTER 9

A STUDY ON THE EFFECT OF ALKYLRESORCINOLS (AR) - METHYLRESORCINOL (MR) AND HEXYLRESORCINOL (HR) ON THE ENZYMATIC ACTIVITY OF TWO HYDROLASES: GLYCOSIDASES (LYSOZYME) AND AMIDOHYDROLASE (PAPAIN) IN THE REACTION OF CHITOSAN HYDROLYSIS

E. I. MARTIROSOVA, N. A. GREBENKINA, and I. G. PLASHCHINA

CONTENTS

9.1 INTRODUCTION

Chitosan is a polyaminosaccharide, a partially deacetylated derivative of chitin. It is widely used in food, biomedical, and chemical industries, as fat blockers, a stabilizer, a preservative for fruits and vegetables, dairy products. From the chemical point of view, the structure of chitosan is a copolymer of glucosamine and N-acetylglucosamine linked by β-1,4-glycosidic bonds. Despite the high biofunctional properties of chitosan, its use is limited because of the high molecular weight and viscosity and, as a result, low absorption *in vivo*. Products of chitosan depolymerization, low molecular weight derivatives as well as chitooligomers exhibit physiological activity greater than chitosan, and therefore have a great potential of application. Low molecular weight chitosans (LMWC) with molecular weight between 5 and 10 kDa shows a strong antibacterial, fungicidal, hypolipidemic, and hypocholesterolemic effect [1–3].

Sphere of chitosan and its low molecular weight derivatives application is constantly expanding. They are used in agriculture as a component of livestock feed, which increases resistance to disease, as part of fertilizers, as a means of prebactericidal seed processing. LMWC can be obtained by physical, chemical, or biological methods. Last method is preferred because it more specific and can be easy controlled by regulation of pH, reaction time, and temperature. Biological methods of chitosan destruction can be devided conditionally to three groups: (1) with using of micromushrooms destructors; (2) with using of purified specific chitolytic enzymes of microorganisms; and (3) with using of another hydrolase classes, nonspecific to given substrate—lipases, glycosidases (lysozyme), and amidases. In the first and second cases, microbial chitinases are used routinely from cultural media directly or separated enzyme preparates. As it was shown, some nonspecific hydrolases are capable to chitosan depolymerization with the same efficiency as chitinases. At the same time, a range of amidhydrolases (pepsin, papain, bromeline, and ficin) can depolymerize chitosan even more effectively than chitinases [4].

Often enzymatic hydrolysis fails to provide a high degree of substratum hydrolysis because of the enzyme inactivation. Rise of the enzymatic activity and functional stability is a crucial problem of current biotechnology. One of the most effective ways of modifying enzymes is using their weak nonspecific interactions with low molecular ligands. Microbial low molecular extracellular metabolites performing the functions of autoinducers of anabiosis represent one of the groups of biologically active substances capable of affecting the activity and stability of lysozyme. These autoregulators represented in a number of bacteria and yeasts by alkylhydroxybenzenes (AHB), alkylresorcinols in particular, induce transition of microbial cells into a hypometabolic (anabiotic) state, and realize this function through interaction with a wide variety of biopolymers of a bacterial cell [5, 6]. Nonspecific influence of these autoregulators on enzymatic proteins is associated with the chemical structure of AHB and the type of their interplay with protein molecules [12].

As was shown earlier, modification of some hydrolytic enzymes by chemical analogs of microbial autoregulatory factors is capable of raising the activity of enzymes in vitro, increasing the depth of hydrolysis of the industrial substrate, and also expanding the temperature and pH-ranges of catalysis [7].

First, part of our work is devoted to researching the chitolytic activity of lysozyme for chitosan hydrolysis. Lysozyme is widely used in medicine and food industries as antimicrobial agent. During last 20 years, lysozyme is intensively used for production baby food and nutrition.

It was shown earlier that efficiency of lysozyme hydrolyzes of different substrates can be increased owing to methylresorcinol, which is the simplest representative of alkylresorcinols. 5-Methylresorcinol has been noted to stimulate the lysozyme activity within the range of concentrations 10^{-7}–10^{-3} M up to 120 percent when peptidoglycane from the Micrococcus luteus is used as a substrate. When nonspecific substrates (colloid chitin, Saccharomyces cerevisiae cells) are used, then the growth of hydrolytic activity was 470 and 400 percent, correspondingly [8]. So, MR shows the ability to change a substrate specificity which is appeared in increasing of hydrolysis rate of the bounds nonspecifically for this enzyme [8]. In our work, the effect of MR on the lysozyme chitolytic activity is shown with using of homogeneous substrate chitosan.

Another part of the work is to study the effect of AR on the chitolytic activity of papain. Papain is a potent proteolytic enzyme of plant origin, belonging to the family of cysteine proteases. Its enzymatic and physiological properties are the subject of series studies, because it plays an important role in the physiology of plants and it is widely used in the food and pharmaceutical industries. In particular, the papain used in the food industry for the softening of meat, production of hydrolysates, clarification of juices and beer, extraction of color and odor plants components, in the dairy industry for cheese manufacture.

The aim of this work was to study the effects of alkylresorcinols (methylresorcinol and hexylresorcinol) on the enzymatic activity of two types of hydrolases such as glycosidases (lysozyme) and amidhydrolases (papain) for chitosan hydrolysis.

9.2 MATERIALS AND METHODS

A sample of hen egg white lysozyme (Sigma-Aldrich, USA) with molecular mass 14,445 g/mol, papain from Papaya carica (Merk, USA) with molecular mass 23,500 g/mol were used. Alkyl-substituted hydroxybenzenes, 5-methylresorcinol monohydrate (5-Methylbenzene-1,3-diol) (Sigma-Aldrich, USA) with molecular mass 142 g/mol and 4-hexylresorcinol (Sigma, USA) with molecular mass 194 g/mol were taken. Substrate for both enzymes was chitosan with deacetylation degree 58 ± 3 percent. It was prepared from commercial chitosan (Sigma-Aldrich) with deacetylation degree 85 percent.

9.2.1 SOLUTION PREPARATION

Protein powder was dissolved in buffer for 2 h and then centrifuged at 20,000 g (Beckman 21, Germany) during 1 h at room temperature. The operating solutions were prepared by mixing equal volumes of protein and alkylresorcinol solutions within 40–50 min just before using in experiment (preincubation time).

Methylresorcinol was dissolved in pure 0.05 M acetic buffer. Hexylresorcinol was dissolved in ethanol solution and then 0.05 M acetic buffer was added to final alcohol concentration 10%. In the control systems, an equivalent amount of the solvent was used instead of the AR solution.

9.2.2 DETERMINATION OF ENZYMATIC ACTIVITY

Chitosan solution (1%; dissolved in 0.05 M acetic buffer, pH 4.5 for lysozyme and 5.0 for papain) was separately treated with modified by AR enzymes (i.e., papain and lysozyme) in the ratio 15:1 (w/w) for lysozyme and 25:1 (w/w) for papain, incubated for 3–4 h at temperatures 37–45°C, followed by arresting the reaction by heat-denaturing the enzyme (100°C, 10 min) and adjusting the pH to 10.0 using 2 M NaOH. The supernatant was separated by centrifugation (10,000 g and 15 min). The concentration of acetylglucosamine, the product of chitosan hydrolysis, was determined in supernatant by the procedure with dinitrosalicylic acid (DNSA) [9].

9.3 RESULT AND DISCUSSION

9.3.1 *EFFECT OF ALKYLRESORCINOLS ON THE LYSOZYME ACTIVITY*

It was established that the use as methylresorcinol (MR), and hexylresorcinol (HR) let to increase the lysozyme activity in the whole concentration range of the used modifiers.

The curve of lysozyme activity (% of control) as a function of the MR concentration has a bimodal shape (Figure 9.1), which is typical for many hydrolases [7]. The first peak corresponds to a MR concentration 0.1 mg/ml and was 60 percent. The second part of the curve has the shape of a curve with saturation.

Maximal effect of lysozyme activity enhancing was attained at a MR concentration 2.0 mg/ml (100%). It can be assumed that the presence of the second peak of lysozyme activity in the chitosan hydrolysis reaction is due to the interaction of lysozyme with ligand molecules in partially associated form. Earlier using methods of isothermal microcalorimetry and dynamic tensiometry it was established the fact of MR self-organization in solution. The value of micelle formation concentration or self-organization (conditionally, CMC) at pH 6.0 is 2.36 mg/ml [10, 11]. Furthermore, the synergistic effect of interaction the MR with lysozyme in solution, resulting in displacement of CMC_{MR} in the protein presence was found using a method of dynamic tensiometry. Clarification of CMC_{MR} values depending of pH in the absence and presence of lysozyme is currently produced.

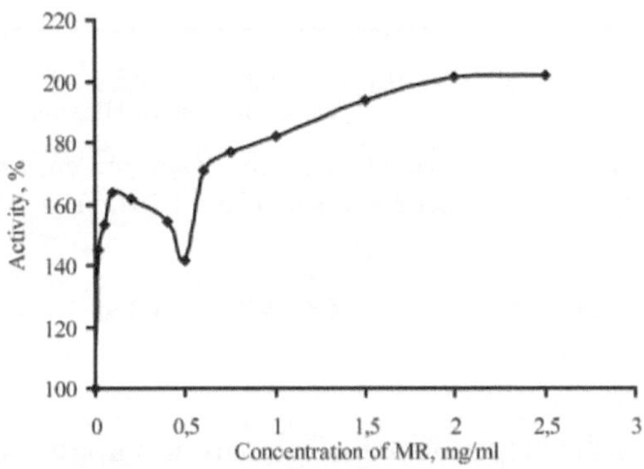

FIGURE 9.1 Effect of MR concentration on the lysozyme enzymatic activity in the process of chitosan hydrolysis. Experiment parameters: 37°C, pH 4.5, 3 h, E/S = 1/15.

In the case of HR, which has more alkyl chain length, hence the higher hydrophobicity, the dependence of the activity from HR concentration has a form of a curve with saturation (Figure 9.2). Apparently, the effect is due, as in the MR case, the transition of the molecular form to HR micelle one which restricts access of modifier into the active site of the enzyme. In this case, the maximum effect was 55 percent of control at a HR concentration of 1.12 mg/ml.

Earlier it was shown that HR has a surface activity. In 5 percent ethanol solution, its CMC is 0.9 mg/ml [5, 6]. In this experiment, saturation region was at HR concentrations 0.7–1.0 mg/ml (in preincubation stage), apparently owing to the presence not only the molecular but micelle form of HR in the solution. The micelle form can serve as a barrier for penetration of the HR in the enzyme active center, and therefore do not provide additional stimulatory action to the enzyme activity.

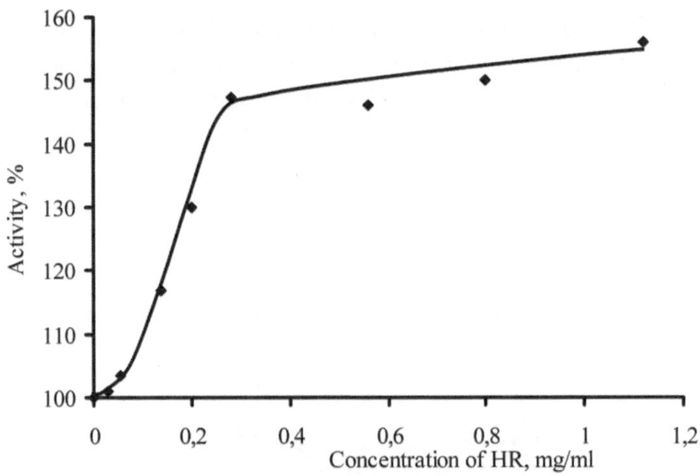

FIGURE 9.2 Effect of HR concentration on the lysozyme enzymatic activity in the process of chitosan hydrolysis. Experiment parameters: 37°C, pH 4.5, 3 h, E/S = 1/15.

9.3.2 EFFECT OF ALKYLRESORCINOLS ON THE PAPAIN ACTIVITY

Figure 9.3 shows the dependence of chitolytic papain activity from MR concentration in the hydrolysis reaction of the chitosan.

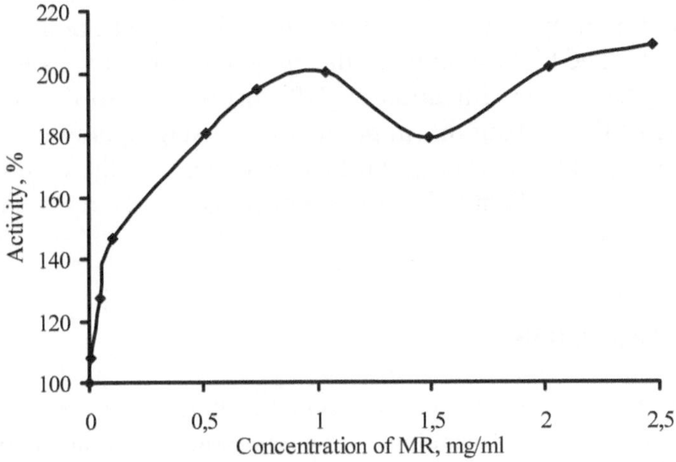

FIGURE 9.3 Effect of MR concentration on the papain enzymatic activity in the process of chitosan hydrolysis. Experiment parameters: 45°C, pH 5.0, 4 h, E/S = 1/25.

Activity curve as in the case of lysozyme has a bimodal character. The position of the first maximum corresponds to the concentration of MR 1.0 mg/ml, which is 10 times higher than for lysozyme. The increase of activity in this case was 100 percent. A similar increase the activity value (100%) is observed in the second peak at MR concentration 2.0–2.5 mg/ml.

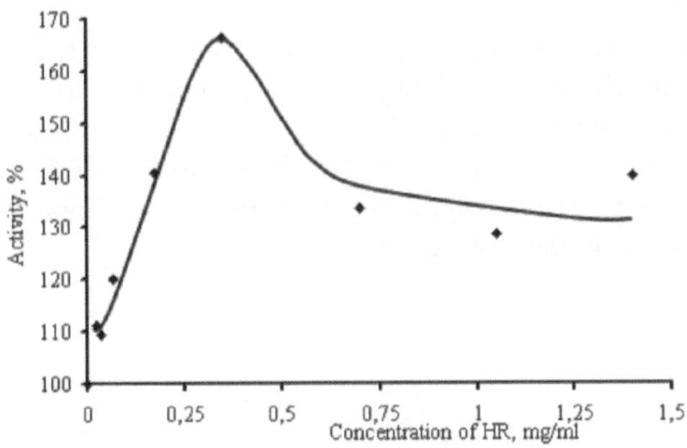

FIGURE 9.4 Effect of HR concentration on the papain enzymatic activity in the process of chitosan hydrolysis. Experiment parameters: 45°C, pH 5.0, 4 h, E/S = 1/25.

HR application for papain modification led to increase enzyme activity maximum up to 65 percent at a ligand concentration 0.3 mg/ml (Figure 9.4). The highest value of the enzymatic activity, as in the case with the lysozyme modification by HR corresponds to the concentration range of 0.7–1.0 mg/ml (in preincubation stage), the limit value of which corresponds to CMC_{HR}. Further increasing of the modifier concentration in the medium led to a slight decrease effect of the enzymatic activity.

9.4 CONCLUSION

The performed experiments provide evidence in favor of the ability of alkylresorcinols—chemical analogs of the homologs of autoinducers of bacterial anabiosis to modify the functional activity of lysozyme and papain and to regulate the efficiency of chitosan hydrolysis. The efficiency of the enzymes modification has been shown to depend on the AR concentration and its structure. The obtained results substantially conform to the data demonstrating the effects of AR on other monosubunit enzymes (trypsin, α-, β-, and glycoamylase, ribonuclease, etc.) and confirm the capability of AR to modify enzymatic proteins nonspecifically to their structure [7, 12]. Changes in the activity of the enzymes in their complexes with AR can be the result of the increasing intramolecular dynamic of enzyme–AR complex, reflected in the enhancing of the equilibrium fluctuations amplitudes [13] and in the decreasing of thermodynamic stability of the protein molecule [14].

Shape of enzyme activity dependence curve from AR concentration correlates with AR self-organizing conditions in the solution. This question requires of additional research.

The experiments have demonstrated the ability of AR to enhance the functional activity of hydrolytic enzymes and thus to increase the efficiency of enzyme in the chitosan hydrolysis.

KEYWORDS

- **Alkylresorcinols**
- **Chitosan**
- **Hexylresorcinol**
- **Lysozyme**
- **Methylresorcinol**
- **Papain**

REFERENCES

1. Vishu Kumar, A. B.; and Tharanathan, R. N.; A comparative study on depolymerization of chitosan by proteolytic enzymes. Carbohydrate Polym. **2004,** 58(*3*), 275–283.
2. Vishu Kumar, A. B.; Varadaraj, M. C.; Gowda, L. R.; and Tharanathan, R. N.; Low molecular weight chitosans—Preparation with the aid of pronase, characterization and their bactericidal activity towards Bacillus cereus and Escherichia coli. Biochimica et Biophysica Acta. **2007,** 1770, 495–505.
3. Shih-Bin Lin, Yi-Chun Lin, and Hui-Huang Chen; Low molecular weight chitosan prepared with the aid of cellulase, lysozyme and chitinase: Characterization and antibacterial activity. *Food Chem.* **2009,** *116,* 47–53.
4. Frolov, V. G.; Dyshkova, Z. G.; and Cherkasova, E. I.; Investigation of chytolytic activity of papain. Materials of 8th International conference Modern prospects in chytin and chytosan investigation. Kazan. June **2006,** 12–17, 315–318 p. (in Russian)
5. Bespalov, M. M.; et al. The role of microbial dormancy autoinducers in metabolism blockade. *Microbiology.* **2000,** *69(2),* 217–223.
6. Kolpakov, A. I.; et al. Stabilization of Enzymes by Dormancy Autoinducers as a Possible Mechanism of Resistance of Resting Microbial Forms. Microbiology. **2000,** *69(2),* 180–185.
7. Martirosova, E. I.; Karpekina, T. A.; and El'-Registan, G. I.; Enzyme modification by natural chemical chaperons of microorganisms. *Microbiology.* **2004,** *73(5),* 609–615.
8. Petrovskii, A. S.; et al. Regulation of the functional activity of lysozyme by alkylhydroxybenzenes. *Microbiology.* **2009,** *78(2),* 146–155.
9. Miller, G. L.; Use of dinitrosalicylic acid reagent for determination of reducing sugar. *Anal. Chem.* **1959,** *31(5),* 426–428.
10. Martirosova, E. I.; Regulation of hydrolase catalytic activity by alkylhydroxybenzenes: thermodynamics of C_7-AHB and hen egg white lysozyme interaction. Biotechnology and the Ecology of Big Cities. New-York, USA: Nova Science Publishers; **2011,** 105–113.

11. Martirosova, E. I.; and Plashchina, I. G.; Improvement of the functional properties of lysozyme by interaction with 5-methylresorcinol. Pharmaceutical and medical biotechnology. New perspectives. New-York, USA: Nova Science Publishers; **2013,** 45–54.
12. Nikolaev, Yu. A.; et al. Changes in physicochemical properties of proteins, caused by modification with alkylhydroxybenzenes. *Appl. Biochem. Microbiol.* **2008,** *44(2),* 143–150.
13. Krupyanskii, Y. F.; et al. Influence of chemical chaperones on the properties of lysozyme and the reaction center protein from Rhodobacter sphaeroides. *Biophysics.* **2011,** *56(1),* 8–23.
14. Plashchina, I. G.; Zhuravleva, I. L.; Martirosova, E. I.; Petrovskii, A. S.; Loiko, N. G.; Nikolaev, Yu. A.; and El'-Registan, G. I.; Effect of Methylresorcinol on the Catalytic Activity and Thermostability of Hen Egg White Lysozyme. Biotechnology, Biodegradation, Water and Foodstuffs. New-York, USA: Nova Science Publishers; **2009,** 45–57 p.

CHAPTER 10

A NOTE ON DISPOSAL OF LIPID COMPOUNDS IN WASTEWATERS

M. S. CHIRIKOVA, T. P. SHAKUN, and A. S. SAMSONOVA

CONTENTS

10.1 INTRODUCTION

Nowadays pollution of aquatic systems with effluents containing various contaminants, including waste lipids has turned into priority problem. Organic components discharged into water reservoirs create favorable conditions for activities of pathogenic bacteria, fungi, and protozoa. They undergo complex biochemical transformations causing thereby secondary contamination and direct adverse effect on local biota [1].

Recovery of municipal sewage and production effluents became an acute challenge. Deregulated urban and industrial runoff interferes with performance of biological decontamination stations based on activated sludge technology. Microbial components of sludge biocenosis are not able to cope with elevated concentrations of the pollutants [2].

The functioning of aeration ponds is complicated by presence in water of huge amounts of lipid wastes separated by mechanical method. This, in turn, arouses the problem of burying the collected by-lipids [3].

In recent years, biopreparations composed of microbial degraders are actively applied for treatment of lipid-saturated wastewaters [4]. Microorganisms synthesize the lipolytic enzymes effectively by breaking down the organic pollutants. Fats and oils are converted into ecologically harmless products [5].

Over 80 dairy plants are operating in Belarus, with overall output of lipidic effluents reaching 10–12 mton/year. Lack of home-made biopreparations intensifying recovery of such wastewaters leads in some cases to illicit dumping and massive environmental damage.

Aim of study was to evaluate application efficiency of microbial preparation promoting bioremediation of wastewaters polluted with lipid compounds.

10.2 OBJECTS AND METHODS OF INVESTIGATION

Microbial preparation Antoil developed at the Institute of Microbiology, National Academy of Sciences, Belarus was chosen as the object of studies. The product incorporates active microbial strains-lipid degraders: *Rhodococcus* sp. R1-3FN, *Rhodococcus ruber* 2B, *Bacillus subtilis* 6/2-APF1, and *Pseudomonas putida* 10AP.

Lipase activity was assayed by Ota-Yamada method [6]. It evaluates alkali-titrated fatty acids released by lipase action from olive oil substrate. The difference between titration results of test and control samples corresponds to the amount of 0.05 n NaOH solution spent for neutralization of fatty acids produced by lipase in the course of olive oil treatment.

Lipolytic activity of microbial cultures LC (u/g) is calculated as follows:

$$LC = \frac{(A.T - 50)}{B}, \quad (1)$$

where *LC* is the lipolytic activity (u/g), *A* is the difference between titration results in test and control samples (cm^3), *T* is the alkali titer, and *B* is the concentration of enzyme solution sample (g/cm^3).

Degradation of lipid substances was studied in 500 ml Erlenmeyer flasks containing 100 ml of mineral medium E8 supplemented with 0.1 percent lipid substrates as nutrition sources (lard, milk fat, sunflower, and olive oil). Aerobic fermentation was carried out on orbital shaker at 150 rpm and temperature 28 °C. Each flask was seeded with 10 percent inoculum (v/v).

Lipid amount in the sample was measured gravimetrically [7].

Fats, oils, other lipids, crude oil fractions were recovered from wastewater by multiple petroleum ether extraction. The resulting extract was divided into two parts.

In one portion, the solvent was evaporated and the residue was weighed to find the total content of lipids and nonvolatile petroleum products according to the formula: 2:

$$x1 = \frac{(m1 - m2) \cdot V2 \cdot 1000}{V1 \cdot V} \quad (2)$$

where *x1* is the total concentration of substances extracted with petroleum ether (mg/l), *m1* is the mass of weighing bottle with residue remaining after evaporation of extractive agent (mg), *m2* is the net bottle weight (mg), *V1* is the extract aliquot volume (ml), *V2* is the volume of graduated flask with extract (ml), and *V* is the volume of analyzed sample (ml).

The other portion of the extract was passed through aluminum oxide, and the content of petroleum products was assessed gravimetrically by (2). The difference in two values constituted actual concentration of fats and other lipids extracted with petroleum ether.

Large-scale mixed culture of bacterial strains *Bacillus subtilis* 6/2-APF1 and *Pseudomonas putida* 10AP was conducted in 300 L fermentor LiFlus SP on Meynell nutrient medium at temperature 28±2 °C during 48 h. Mixed fermentation of bacterial strains *Rhodococcus sp.* R1-3FN and *Rhodococcus ruber* 2B was performed in fermentor LiFlus SP (300 l) on Meynell medium at temperature 28±2 °C during 98 h.

Chemical oxygen demand was determined by express method [7].

10.3 RESULTS AND DISCUSSION

Screening of microorganisms for the ability to utilize fats and oils as nutrition sources embraced 40 cultures collected by researchers from laboratory of xenobiotic degradation and bioremediation of natural and industrial media. Collection entries were isolated from municipal sewage and activated sludge of decontamination stations treating wastewaters of organic synthesis and dairy/meat processing overloaded with fatty substances.

Degrading potential of tested microbial species was judged by the ability to grow on synthetic medium E8 containing 1 percent concentration of lipids supplied from vegetable oils—sunflower, olive, and animal fats—pork and milk.

Overwhelming majority of cultures capable to grow on media with oils and fats is represented by genus *Rhodococcus*. It accounted for 67.5 percent among 40 examined isolates, 20 percent share belongs to *Bacillus* genus, and the rest 12.5 percent is occupied by *Pseudomonas* genus.

Ten cultures displayed most active growth, with four strains taking the dominant lead: *Rhodococcus* sp. R1-3FN, *Rhodococcus ruber*2B, *Bacillus subtilis* 6/2-APF1 and *Pseudomonas putida* 10AP.

Maximal lipolytic activity of microorganisms consuming lipids as a sole source of nutrition revealed in cultural liquid of selected superactive variants varied from 0.65 to 0.70 U/mg. Strains with inferior lipolytic activity showed the values in the range 0.3–0.6 U/mg (Table 10.1).

TABLE 10.1 Lipase activity of tested microorganisms

S. No.	Strains	Specific Activity, u/mg Protein
1	*Bacillus subtilis* 6/2-APF1	0.68 ± 0,04
2	*Pseudomonas putida* 10AP	0.65 ± 0,04
3	*Pseudomonas fluorescens* 12B	0.30 ± 0,05
4	*Rhodococcus sp.* R1-3FN	0.72 ± 0,04
5	*Rhodococcus erythropolis* 70F	0.51 ± 0,06
6	*Rhodococcus erythropolis* 23F	0.50 ± 0,05
7	*Rhodococcus opacus* 100B	0.47 ± 0,03
8	*Rhodococcus opacus* 29D	0.52 ± 0,04
9	*Rhodococcus ruber* 1B	0.60 ± 0,04
10	*Rhodococcus ruber* 2B	0.70 ± 0,03

Strains *Rhodococcus* sp. R1-3FN, *Rhodococcus ruber* 2B, *Bacillus subtilis* 6/2-APF1 and *Pseudomonas putida* 10AP were rated as the most promising ingredients of microbial preparation elaborated to accelerate removal of lipid pollutants from wastewaters.

Application prospects of aforementioned strains in biotechnology of formulating remediation preparation are grounded on the ability of pure microbial cultures to use fats and oils as nutrition sources.

Decomposing activity of four selected microbial strains was evaluated in 500 ml Erlenmeyer flasks where 0.1 percent concentrations of lard, milk fat, olive, and sunflower oil served as nutrient substrates. Fermentation proceeded for 7 days on orbital shaker at agitation rate 150 rpm. Upon 7 days residual lipid substances were determined in the cultural liquid (Table 10.2).

TABLE 10.2 Oil and fat degradation efficiency of Antoil microbial constituents

Strain	Decomposition Degree (%)			
	Lard	Milk Fat	Sunflower Oil	Olive Oil
Rhodococcus ruber 2B	84.3	88.2	86.2	84.2
Rhodococcus sp. R1-3FN	90.2	90.9	82.4	88.2
Pseudomonas putida 10AP	91.9	89.2	86.8	87.4
Bacillus subtilis 6/2-APF1	92.5	94.2	86.1	90.4

Investigation of the ability of tested microbial variants to digest oils and fats as nutrients demonstrated that all four strains catabolized both types of substrates. Strains *Rhodococcus* sp. R1-3FN, *Bacillus subtilis* 6/2-APF1 preferred lard, and milk fat. Strain *Pseudomonas putida 10 AP* found lard most appetizing. Utilization efficiency of lard, milk fat, sunflower and olive oil by *Rhodococcus ruber 2B* constituted 84.3 percent, 88.2 percent, 86.2 percent, and 84.2 percent, respectively.

Industrial trials of Antoil performance were completed at decontamination unit of Kopyl dairy plant and at biological sewage processing facilities of Kopyl municipal public utility network.

Antoil preparation was fed into four segments of aeration tank treating effluents of Kopyl dairy plant in amount 100 L per each tunnel and runoff was ceased afterwards. Biopreparation was incubated with wastewaters at continuous aeration during 8 h.

Initial contamination level of effluents denoted as COD in four sectors of aeration tank prior to Antoil supply equaled 840, 860, 830, 850 mg O_2/L, respectively.

After Antoil treatment COD values sharply declined to 440, 430, 400, 410 mg O_2/L, respectively. Vital COD indices fell by 47.6 percent, 50 percent, 51.8 percent, 51.7 percent, respectively, making the average percentage 50.2 percent.

Concentration of fatty substances in effluents of Kopyl dairy plant before Antoil application reached 750 mg/L. In post-Antoil samples, concentration of lipid pollutants was drastically reduced by 90.4 percent to 72 mg/L. The obtained data are summarized in Table 10.3.

TABLE 10.3 Results of Antoil efficiency trials at decontamination facilities of Kopyl dairy plant

Experimental Variant	COD Values (mg O_2/l)				Amount of Lipid Substances (mg/l)
	1-t Segment of Aeration Tank	2-d Segment of Aeration Tank	3-d Segment of Aeration Tank	4-th Segment of Aeration Tank	
Before Antoil application	840	860	830	850	750
After Antoil treatment	440	430	400	410	72
Reduction of parameters (%)	47.6	50	51.8	51.7	90.4

Decrease of pollution level of dairy effluents illustrated by 50.2 percent COD decline and decay of lipid substances by 90.4 percent laid the basis for further series of Antoil performance tests at biological detoxification station of Kopyl public utility network.

Successful testing program demonstrated that introduction of defatting preparation into aeration tank resulted in 87.6 percent reduction of COD level in the treated wastewaters. Initial amount of lipid pollutants in effluents (48.8 mg/L) fell drastically to trace concentrations, that is, decontamination efficiency reached 99.9 percent.

10.4 CONCLUSIONS

High degrading activity of Antoil microbial constituents (84.3–92.5%) and efficiency of wastewater recovery in terms of COD reduction (87.6%) and lipid decomposition (99.9%) registered during industrial trials at Kopyl municipal decontamination station evidence its attractive application prospects for intensification of lipid disposal in effluents. Technological introduction of this microbial preparation into bioremediation schemes will be of considerable social significance, because it will promote degradation of lipid pollutants in the discharged wastewaters and hence will improve ecological situation in Belarus.

KEYWORDS

- **Chemical oxygen demand**
- **Disposal efficiency**
- **Intensification of decontamination**
- **Lipid substances**
- **Lipolytic activity**
- **Microbial degraders**
- **Microbial preparation**

REFERENCES

1. Ivchatov A. L.; and Glyadenov, S. N.; Another aspect of biological decontamination of effluents. *Ecol. Ind. Russ.* 2003, 4, 37–40, (in Russian)
2. Samsonova, A. S.; Glushen, E. M.; Chirikova, M. S.; and Petrova, G. M.; Microorganisms intensifying disposal of lipid substances in wastewaters. Microbial Biotechnologies: Basic and Applied Aspects. Minsk: Belaruskaya navuka Press; 2012, 250–259 p. (in Russian)
3. Poskryakova, N. V.; Development of Biopreparation Basis for Lipid Degradation. Ufa, 2007, 24 p. (in Russian)
4. Murzakov, B. G.; Zaikina, A.I.; Zobnina, V. P.; Listov, E. L.; Zorina, L. V.; and Rogacheva, R. A.; Biotechnological method for removal of fats and oils from wastewaters. Russian Patent No. 2161595, 2001, 23 p. (in Russian)
5. Reimann, J.; and Gotsche, A.; Reinigung fetthaltiger Abwasser der Frostfischindustrie nut thermophilen Mikroorganismen. *Chem. Ind. Techn.* 2002, *5,* 634 s.
6. Gerhard et al. eds. Methods of General Bacteriology. Mir, Moscow, 1983, 536 p. (in Russian)
7. Lurie, Yu. Yu.; Analytical chemistry of industrial effluents. Chemistry Series. Moscow; 1984, 448 p. (in Russian)

CHAPTER 11

INCREASING OF THE ROAD PAVEMENT'S LIFETIME BY INTRODUCING OF SUPERFINE ELASTOMER PARTICLES IN THE BITUMINOUS BINDER IN SUCH PAVEMENTS

N. I. CHEKUNAEV and A. M. KAPLAN

CONTENTS

11.1 INTRODUCTION

The constant interest in elucidation of new possibilities to increase the durability of asphalt concrete pavements continues to persist for several decades. Such pavements are composites consisting of gravel, sand, and the polymer-containing composite (bitumen) in rationally chosen ratios. According to modern notions, it is assumed that the durability of asphalt concrete pavements is determined by the appearance of numerous major trunk cracks in the pavement. This conclusion is confirmed, in particular, in numerous reports (over 130), presented at a special conference devoted exclusively to the problem of cracking in pavements [2]. It was shown earlier and in reports of the conference that a certain increase in the durability of pavements can be achieved by introducing synthetic rubbers or crushed technical rubber in the bituminous binder of elastomers. However, very important question about a quantitative determination of the optimum size of elastomers particles introduced into the bitumen for providing of maximum durability of pavements, until recently, remained open. The theoretical solution of above problem is presented below.

11.2 DEVELOPMENT OF CRACKS IN CYCLICALLY LOADED POLYMER-CONTAINING COMPOSITES

The main destruction mechanism of solids is the nucleation and growth of cracks. One of the most effective ways to improve the performance properties of the material is the introduction into bitumen an elastic modifier such as SBS or rubber in an amount of 1–10 percent. Asphalt concrete (AC) is a combination of different mineral components mixed with bitumen. Bitumen is a small portion of asphalt material, but plays a very important role in binding together the material, wherein the cracks are generally propagating in a bituminous component.

In this paper the negative Celsius temperatures were assumed, when the material can be regarded as elastic solid with sufficient accuracy. In present paper classical notions of Griffiths and Orowan were used. These notions, suitable for description of the cracks behavior in homogeneous materials, were modified for description of crack development in heterogeneous materials. In heterogeneous materials, such as AC, the embryonic voids with the characteristic sizes a_{ini} 20–50 nm have a concentration c_0 =

$10^{17}-10^{21}$ м$^{-3}$ [3]. The volume fraction occupied by these voids is small (~10^{-5}–10^{-4} of the total volume) however, their effect on the strength and durability of the materials can be decisive.

The mechanical impact on heterogeneous materials leads to existence of areas with high local stresses $\sigma_{loc} = \beta_1 \cdot \sigma_0 \approx (3 \div 10)\sigma_0$ [4], where β_1 is the overstress factor. Areas with the most elevated local stress, comparably with the average stress $\sigma_0 \sim 1.5$ MPa, are located in asphalt concrete near the acuminated borders of mineral component particles in narrow interlayers of binder between the mineral particles of the asphalt. Additional stress concentrators are available in asphalt concrete macrocracks, original, and also the newly formed under intense mechanical stresses. The local stresses σ_{loc} in areas of hetero-structures spaced at distance r from a crack's tip exceed external stress in such structures by the factor $\beta_2 = \sqrt{a/r}$. The estimation gives $\beta_2 \sim 50$ for the region with the size of 5 microns spaced apart from the tip of the macrocrack with length $a = 1$ cm. Evaluation of the resultant stress near the crack's tip in the vicinity of the border of mineral particles gives the value $\sigma_{loc} = \beta_2 \cdot \beta_1 \cdot \sigma_0 > 200$ MPa. These high local stresses up to 300 MPa, as noted in [5], were found also in the polymeric material. It should be noted that for larger local stress smaller spatial scale is typical. Regarding ultrahigh local stress of a few hundreds of MPa, the size of the corresponding area is between one and several micrometers. It is in these areas the nucleation and growth of cracks occurs, with their transformation into supercritical locally.

Rubber modifier is characterized by a slow rate of fatigue growth under external cyclic loads as well as high energy γ_R of new surface generation, significantly exceeding (by tens or hundreds times) bitumen surface energy γ_B. Therefore the crack critical size in the rubber is significantly larger than the critical size in the bitumen $a_{cr}^{(R)} >> a_{cr}^{(B)}$.

An important role in increasing the strength of polymers play "crazes" formed near modifier particles at adding modifier. (Refer for example, [6, 7]). Energy expenses for the microcracks formation allow fast enough to reduce the level of tension of the polymer sample and thereby hinder its destruction. Also, the presence of crazes in the vicinity of the modifier particles leads to the fact that more amount of energy is needed to create a new surface. The principle of retardation and cessation (arrest) of fast propagating cracks in mechanically loaded AC pavement by particles of elastic modifier is demonstrated in Figure 11.1.

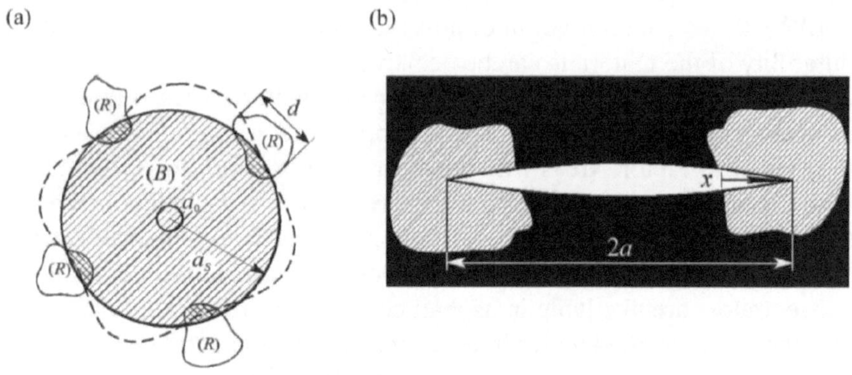

FIGURE 11.1 Schematic diagram of breaking and arrest of the disk-shaped crack at junction with elastic modifier particles in bitumen. (a) The cross section of modified bitumen sample by the plane coinciding with plane of the disk-shaped crack (shaded). For additional explanations refer the text. (b) The cross section of the same sample by the plane of perpendicular to the crack plane (light color). Crack is surrounded with bitumen (shown in black) and partially penetrates into elastic modifier particles (shaded).

Initial microvoids have completely different shapes. Some portion (α of them) have a shape, which can be approximately regarded as submicro-cracks with the size a_{ini}. For some time τ they grow up to sufficiently small critical size $a_{cr}^{(B)}$ in the bitumen matrix. These cracks are disk-shaped, so the size of the crack can be characterized by its radius a. After exceeding the critical size rapid cracks growth starts in the asphalt until it meets an obstacle in the form of a particle modifier. And, as shown in Figure 11.1, losing energy crack is arrested by modifier particles. Fatigue crack growth continues, albeit relatively slow. Cracks grow relatively fast in the bitumen and much slower when penetrating into the rubber. Cutting of elastic modifier particles by the trapped crack due to fatigue growth requires a long time. Critical size of the disk-shaped crack is determined from the equation [8]:

$$a_{cr} = 1.6E\gamma/\sigma^2 \tag{11.1}$$

where E is modulus of elasticity of the material. We estimate the value of the critical radius in the various components. Values of E_B and γ_B depend on the bitumen grade and the ambient temperature. To estimate we took intermediate values E_B = 1 GPa and γ_B = 10 J/m² [9], E_R = 100 MPa and

$\gamma_R = 20$ kJ/m^2 [10]. Then, when $\sigma_{loc} = 250$ MPa, the critical radius of the disk-shaped cracks in the bitumen is equal to $a_{cr}^{(B)} = 250$ nm and in rubber $a_{cr}^{(R)} = 50$ microns. Grown to a size greater than $a_{cr}^{(B)}$, the crack can quickly accelerate and turn into a destructive trunk crack in the pure bitumen. To prevent such crack development the modifier rubber particles are incorporated in the bitumen. Let us give here briefly the theory of crack's cessation in solids by modifier particles [1, 11]. The total energy of a disk-shaped crack of radius a:

$$a^3\left[v^2/v_m^2 - \left(1 + a_{cr}^{(B)}/(2a)\right)\left(1 - a_{cr}^{(B)}/a\right)^2\right] + 3a_{cr}^{(R)}A/(2\pi) = 0\,,\quad v_m^2 = cE/(k\rho) \qquad (11.2)$$

where ρ is the material density, v is the rate of crack propagation, v_m is the limiting crack velocity, A is a new generated area at crack penetration into the rubber component. Critical crack size $a_{cr}^{(B)}$ in the bitumen component is determined from the condition $dE_{tot}\,(a, v = 0)/da = 0$. When crack is crossing inclusions, the last term on the left side in Eq. (11.2) plays a significant role . Approximately, in accordance with Figure 11.1, one can take $A = z\pi xd/6$, where x is an average path passed by the crack in the rubber inclusion, z 4 is the coordination number, approximately equal to the minimum number of inclusions required to stop crack. Coefficient of d is chosen so that at the full dissecting of inclusion particle ($x = d$) the formed area was equal to the average area of particles cross-section. From the Eq. (2) we can find a mean path x_S, passed by the crack in the modifier particle before crack stops. At the condition $a_{cr}^{(B)} << a_S$, solving the Eq. (11.2), we obtain:

$$x_S = 4a_S^3/\left(za_{cr}^{(R)}d\right) \qquad (11.3)$$

where a_S is crack radius at which it is stopped. To stop the crack the inequality $x_S < d$ is necessary and we have the inequality:

$$d^2 > 4a_{stop}^3/\left(za_{cr}^{(R)}\right) \qquad (11.4)$$

The crack's circular front when pinned is strongly curved [12]. If the crack front deviation is $\delta \times a$, then the active volume at the crack passage is equal to $2\pi\delta a_S^2 d$, then the number of inclusions, caught in this volume equals $2\pi\delta da_S^2 C_m = z$, $C_m = 3r_m^{-3}/4\pi$, where r_m is the mean distance between modifier particles, C_m is their concentration. Since the fractional volume is equal $V_{fr} = R^3/r_m^{\ 3} = d^3/8r_m^{\ 3}$, we obtain $a_S = z^{1/2}d(12\delta V_{fr})^{-1/2}$. Substituting this value in Eq. (11.4), we have $d < 0.25\ (12\delta)^{3/2}\ z^{-1/2} V_{fr}^{\ 3/2}\ a_{cr}^{\ (R)} = d_{max}$. For self-consistency, this condition coincides with the condition $a_S < a_{cr}^{\ (\mathrm{av})}$ it is required to put $\delta = 1/3$, that is, cracks exceeding the critical size in the bitumen $a_{cr}^{\ (B)}$ will grow rapidly to the size of $a_S = z^{1/2}d(12\delta V_{fr})^{-1/2}$, until stopped by modifier particles. After this the rapid crack growth turns to the slow fatigue growth from the size a_S to a critical size in averaged medium $a_{cr}^{\ (\mathrm{av})} = V_{fr}a_{cr}^{\ (R)}$. The scheme of crack growth is shown in Figure 11.2.

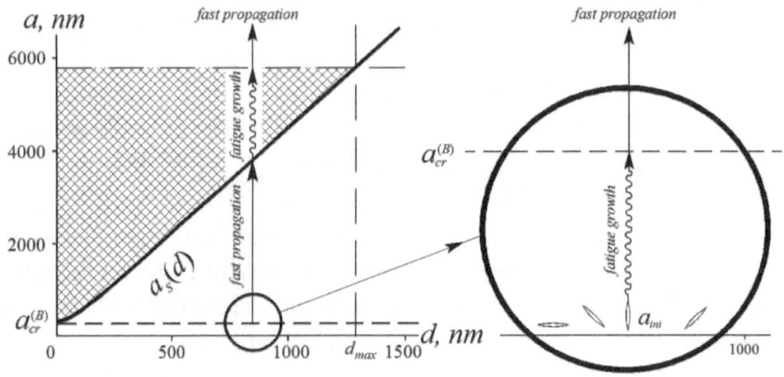

FIGURE 11.2 Scheme of crack growth in rubber-bitumen composite at a given size d of the inclusion particle.

In the calculations the fatigue crack growth as per Paris' law was used (rewritten here in the convenient form):

$$da / dN = C(\Delta K)^n\ ;\ K = (\sigma_{loc}/\sigma_A)\sqrt{a/l_A}\ , \tag{11.5}$$

where C and n are material parameters: amplitude and stress intensity factor change. Fate of crack can be determined as follows. In Eq. (11.5) the chosen constants $\sigma_A = 40$ MPa, $l_A = 100$ nm. The initial embryonic submicrocracks of size a_{ini} grow slowly and then exceed the size $a_{cr}^{\ (B)}$. Thereafter, they are rapidly supercritically grow in the bitumen to a size $a_S(d)$, until arrested by modifier particles. After stopping by modifier particles cracks will grow by fatigue route to the critical size $a_{cr}^{\ \mathrm{av}}$ of averaged medium. The two parallel process of cracks conversion supercritically are possible: (1) Accumulation of cracks at the boundary $a_S(d)$ and their subsequent merging to generate a supercritical crack. Let the concentration of initial submicrocracks be αc_0. As a result of cyclic impact they transform into supercritical state for bitumen matrix in characteristic time interval τ. Then they propagate fast in the matrix to the size $a_S(d)$ and then stop. Concentration c of the stopped cracks varies as $dc/dt=\alpha c_0/\tau$. At a concentration of $c = c_{max} \sim a_{stop}{}^{-3}$ the arrested cracks begin to touch each other and then merge to generate a large supercritical crack which can turn into a trunk crack, destroying the entire sample. Hence, the time (measured in number of cycles) of transition into larger supercritical crack because of their accumulation and merging, that is, time required for fatigue growth from the size of crack nucleus a_{ini} to the critical size in bitumen $a_{cr}{}^{(B)}$ equals:

$$N_1 = c_{\max}\tau/\alpha c_0 = 8z^{-3/2}V_{fr}^{3/2}\tau/(\alpha c_0 d^3) \tag{11.6}$$

The transition time τ (measured in number of cycles), can be calculated from Eq. (11.7) and equals to $N^{(B)}$ in Eq. (11.8). The value τ can be estimated as durability of the pure bitumen material.

(2) The second process is fatigue growth of arrested crack from the length of $a_S = 0.5z^{1/2}V_{fr}{}^{-1/2}d$ to a length $a_{cr}{}^{(av)}$. Assuming that the fatigue crack growth occurs as per the Paris law (11.5), we find the number N_2 of cycles required to reach the supercritical crack size for the averaged medium:

$$N_2 = N_\infty^{(R)}\left(1-2n_R^{-1}\right)\left(1-\left(a_S/a_{cr}^{\text{av}}\right)^{n_R/2}\right)d/a_S + N_\infty^{(B)}\left[\left(1-\left(a_S/a_{cr}^{\text{av}}\right)^{n_B/2-1}\right)-\left(1-2n_B^{-1}\right)\left(1-\left(a_S/a_{cr}^{\text{av}}\right)^{n_B/2}\right)d/a_S\right]$$

where

$$N_\infty^{(B,R)} = 1/\left[\pi^{n/2}(n_{B,R}/2-1)C_{B,R}a_S^{n_{B,R}/2-1}\sigma^{n_{B,R}}\right] \tag{11.7}$$

Here $C_{B,R}$ and $n_{B,R}$ indicate C_B, n_B for bitumen and C_R, n_R for rubber. Since these two processes are parallel, then their rates (proportional to $N_1{}^{-1}$ and $N_2{}^{-1}$) were added. For obtaining Eq. (11.7) it was taken in account that fatigue growth of the crack takes place partially in rubber particles (portion V_{fr}) and partially in bitumen (portion $1 - V_{fr}$). Then the required number of cycles N to crack transformation into supercritical can be estimated as

$$N^{-1} = N_1^{-1} + N_2^{-1} \tag{11.8}$$

Relative increase of the durability N_r comparable with the case of the unmodified bitumen equals to the quotient $N_r = N/\tau$. The results of calculations are shown in Figure 11.2. Calculations were made with use of following parameters in Paris' law: $C_B = 9.7\times10^{-7}$m/cycle, $n = 1.5$ for bitumen [13] and $C_R = 3\times10^{-9}$ m/cycle, $n = 0.211$ for rubber [14].

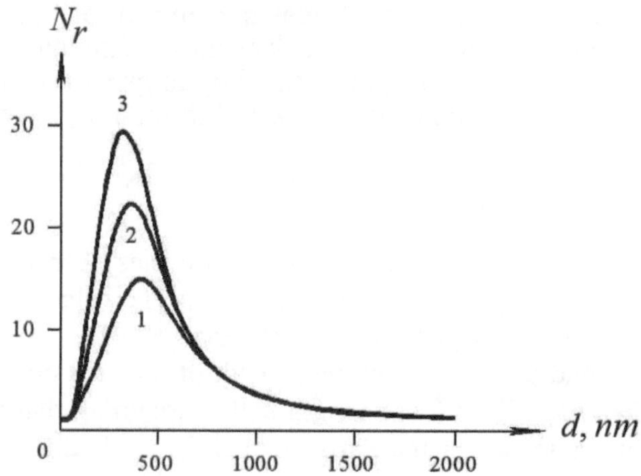

FIGURE 11.3 Calculated dependences of the relative durability N_r of cyclic loaded rubber-bitumen binder samples on the sizes of the rubber particles introduced into the bitumen. Curve 1—at an average mechanical stress in the pavement $\sigma = 2{,}5$ MPa, Curve 2 ($\sigma = 2{,}75$ MPa), Curve 3 ($\sigma = 3$ MPa). The fractional volume of rubber inclusions in rubber-bitumen binder $V_{fr} = 5$ percent.

11.3 CONCLUSION

As can be seen from Figure 11.3, it is necessary to use very small (150–750 nm) rubber particles in the asphalt binder to ensure optimal high durability of pavement. It should be noted that obtaining of the most inexpensive raw materials (tire rubber fine particles) for successful modifying bitumen was possible only by using the original method of "high-temperature shear-induced grinding of polymers and their composites" developed by Prof. V. G. Nikol'skii. Recently, it was shown that the durability of bitumen binder samples manufactured with using rubber particle sizes (about 0.5–0.8 μ) is several times greater the durability of samples manufactured from known domestic and foreign brands of bitumen binders modified by rubber using rubber particle sizes higher than 2–4 μ [15].

ACKNOWLEDGMENTS

The authors thank prof. V. G. Nikol'skii for valuable discussions and prof. G. E. Zaikov for support of this study.

KEYWORDS

- **Elastome—bitumen binder**
- **Micro-cracks**
- **Road pavements**
- **Tire crumb rubber**
- **Trunk cracks**

REFERENCES

1. Kaplan, A. M.; and Chekunaev, N. I.; Theoretical Foundations of Chemical Engineering. **2010,** *44(3),* 339–347.
2. Proceedings "7th RILEM International Conference on Cracking in Pavements". Scarpfs, A.; Kringos, N.; Al-Qadi, I.; and Loizos, A.; eds. Dordrecht, Heidelberg, New York, London: Springer Publisher; **2012,** 1378 pp.
3. Cheremskoy, P. G.; Slezov, V. V.; and Betekhin, V. I.; Pores in Solids. Moscow: Energoatomizdat; **1990,** (in Russian).
4. Rudenskii, A. V.; Road Asphalt Pavements. Moscow: Transport; **1992,** (in Russian).
5. Bucknall, C. B.; Deformation mechanisms in rubber-toughened polymers. In "Polymer Blends". Paul, D. R.; and Bucknall, C. B.; eds. New York, Toronto: John Wiley & Sons Inc.; **2000,** *2,* 83–118 pp.
6. Michler, Georg H.; Electron Microscopy of Polymers. Springer; **2008**.
7. Pearson, R. A.; and Pruitt, L.; Fatigue-crack propagation in polymer blends. In: "Polymer Blends". Paul, D. R.; and Bucknall, C. B.; eds. New York, Toronto: John Wiley & Sons Inc.; **2000,** *2,* 269–300.
8. Elliott, H. A.; An analysis of the conditions for rupture due to griffith cracks. *Proc. Phys. Soc.* **1947,** *58,* 208–223.
9. Hesp, M.; Development of a Fracture Mechanics-Based Asphalt Binder Test Method for Low Temperature Performance Prediction. Final Report for Highway IDEA Project 84. Transportation Research Board of the National Academies. **2004**.
10. Al-Quraishi, A. A.; The Deformation and Fracture Energy of Natural Rubber Under High Strain Rates. Ph.D., University of Akron, **2007**.
11. Chekunaev, N. I.; and Kaplan, A. M.; Acceleration, retardation and crack arrest in stressed heterogeneous structures. *Key Eng. Mater.* **2011,** 462–463, 506–511.

12. Kamaya, M.; Stress intensity factors of surface crack with undulated front. *JSME J. Ser. A.* **2006,** *49,* 529–535 pp.
13. Khalid, H. A.; and Aramendi, I.; Measurement and Effective Evaluation of Crack Growth in Asphalt Mixtures, in the Book Pavement Cracking: Mechanisms, Modeling, Detection, Testing, and Case Histories. Al-Quadi, I. L.; et al. eds. London, UK: Taylor & Francis Group; **2008**.
14. Schubel, P. M.; Gdoutos, E. E.; and Daniel, I. M.; Fatigue characterization of tire rubber. *Theor. Appl. Fract. Mech.* **2004,** *42(2),* 149–154 pp.
15. Nikol'skii, V. G.; Kaplan, A. M.; and Chekunaev, N. I.; et al.; ISSN 1990_7931, *Russ. J. Phys. Chem. B.* **2014,** *8(4),* © Pleiades Publishing, Ltd, 2012. Original Russian Text © **2014,** will be published in Khimicheskaya Fizika, **2014,** *33(7).*

CHAPTER 12

USE OF MICROSIZED FERROCOMPOSITES PARTICLES FOR IMMOBILIZATION OF BIOLOGICALLY ACTIVE COMPOUNDS

LUBOV KH. KOMISSAROVA and VLADIMIR S. FEOFANOV

CONTENTS

12.1 AIMS AND BACKGROUND

Magnetic nano- and microsized particles can be used for various biomedical applications for example, cell separation, immobilization of enzymes and viruses, detoxification of biological liquids, magnetic drug targeting und others [1–5], the most widespread being neutron capture therapy (NCT) when they become compounds with ^{10}B (BNCT). There are two boron containing compounds, one being L-borophenilalanin (L- BPA) used in clinical practice [6]. The aim of the research is to work out new methods to modify the surface of different chemical content microsized ferrocomposites particles with biocompatible materials for the immobilization of biologically active compounds and to evaluate the possibility to use them as sorbents for extracorporal detoxification of patient's blood and conserve donor blood after purification from free hemoglobin and barbiturates by the method of magnetic separation and as carriers for magnetically guided target delivery of L-BPA at BNCT.

12.2 EXPERIMENTAL

We studied the composites, and they were iron-silica ($FeSiO_2$) contains 50 percent Fe, 50 percent SiO_2, iron-carbon (FeC) contains 44 percent Fe, 56 percent C, iron-carbon-silica ($FeCSiO_2$) contains 50 percent Fe, 40 percent C, and 10 percent SiO_2. Iron contains 90 percent restored iron and 10 percent, Fe_3O_4, sized 0.02–0.1 mkm when obtained by plasmochemical method [5], and magnetite (Fe_3O_4) sized 0.1–0.5 mkm when synthesized by chemical method [2]. Diameter of ferrocomposites microparticles in 0.6 percent of albumin solution is 1–2 mkm^2. The powders of ferromagnetics were treated (as suspension in distilled water) with ultrasonic waves (frequency 22 kHz) in order to eliminate aggregation and to attain a homogenous distribution of the particles in suspension. The particles' surface, besides being the same of composites $FeSiO_2$ and $FeCSiO_2$ were biocompatible, and were covered with albumin, or gelatin, or dextran. Carboxilate- magnetic particles were obtained, with bovine albumin or gelatin coating following aldehyde modification. Aldehyde-magnetic particles were obtained with dextran coating following $NaJO_4$ activation. We coated the particles by mixing a suspension of particles with albumin or gelatin or dextran with Mm 67,000 Da (Sigma) using ultrasound, with the following 1 h incubation at 20°C, separation of particles on Sm-Co magnet with inductance of 0.1–0.15 Tl. Thereafter the particles were incubated in the modificator solutions, formaldehyde (Russia), or glutaraldehyde, or $NaJO_4$ (Sigma) and washed with distilled water. Surface-modified particles were kept at 10 percent concentration in a physiological solution.

We had used bovine hemoglobin (Sigma) which contained up to 75 percent methemoglobin and up to 25 percent oxyhemoglobin. Immobilization of hemoglobin and barbiturates (Russia) was carried out with 30 sec incubation with the suspension of particles in physiological solution and in a model biological liquid (0.6% albumin in physiological solution) at 20°C (pH 7.4) at different weight ratios of composite/substance that is at 10, 20, 50, and thereafter the particles were separated on Sm-Co magnet. We had chosen an incubation time 30 sec based on the length of contact of biological liquids with suspended magnetic microparticles in the device for extracorporal detoxification of blood using the method of magnetic separation [1]. Concentrations of compounds in the solutions were measured by differential visual and UV-spectroscopy. The sorption efficiency of ferrocomposites was evaluated as the ratio of the quantity of the adsorbed substance to its initial amount (w/w), expressed in percent and in milligram to grams composite (absorptive capacity) for a certain weight ratio of composite/substance.

Immobilization of L-BPA (Lachema) was carried out with 10 min incubation and with the suspension of particles in acidified water solutions at different weight ratios of composite/L-BPA. The dynamics of L-BPA desorption was studied by incubation of magnetic preparations with immobilized of L-BPA with fresh aliquots of 0.6 percent albumin at 37°C and by following registration of supernatant on absorption UV-spectra. Concentration of desorbed BPA was evaluated on the calibri curve.

12.3 RESULTS AND DISCUSSION

12.3.1 IMMOBILIZATION OF HEMOGLOBIN AND BARBITURATES

The maximal sorption efficiency of hemoglobin on unmodified ferrocomposites particles showed, with Fe_3O_4 and Fe-particles, as 40.0 and 37.8 mg/g respectively. The results of sorption efficiency of modified ferrocomposites particles to hemoglobin are presented in Table 12.1.

TABLE 12.1 Sorption efficiency of gelatin-modified Fe_3O_4-particles to hemoglobin at different weight ratios Fe_3O_4/Hb in physiological solution at pH 7.4

Types of Composites m	Sorption, Average ± SD (%) / Absorptive Capacity, Average ± SD (mg/g)						
	Unmodified		Gelatin-Covered		Gelatin + Glu-tarald − Modified		Gelatin + Formald Modified
Fe_3O_4/Hb	20	50	20	50	20	50	20
	41.6 ± 4.7	70.4 ± 9.3	34.9 ± 4.9	57.8 ± 6.1	26.9 ± 3.6	47.8 ± 5.9	41.0 ± 4.2
	20.8 ± 2.4	14.1 ± 4.7	17.4 ± 2.4	11.6 ± 1.2	13.4 ± 1.8	9.6 ± 1.2	20.5 ± 2.1

Table 12.1 shows that the sorption efficiency of magnetite to hemoglobin after covering the surface with gelatin is decreasing, glutaraldehyde-modification leads to its further decrease. Sorption efficiency of magnetite does not exchange practically after formaldehyde modification with gelatin covered particles. The same character of sorption efficiency of gelatin and albumin—modified particles to hemoglobin is discovered for Fe and FeC particles. Sorption efficiency is decreasing after covering of particles by proteins, and the glutaraldehyde-modification leads to its further decrease and does not change after formaldehyde modification. Immobilization of hemoglobin on aldehyde-modified particles, evidently is because of forming hydrogen connections between carboxilate groups of proteins and amino-groups of hemoglobin. Decrease of hemoglobin adsorption efficiency of particles with glutaraldehyde-modified surface is predetermined, obviously, because of stereochemical factor (Figure 12.1).

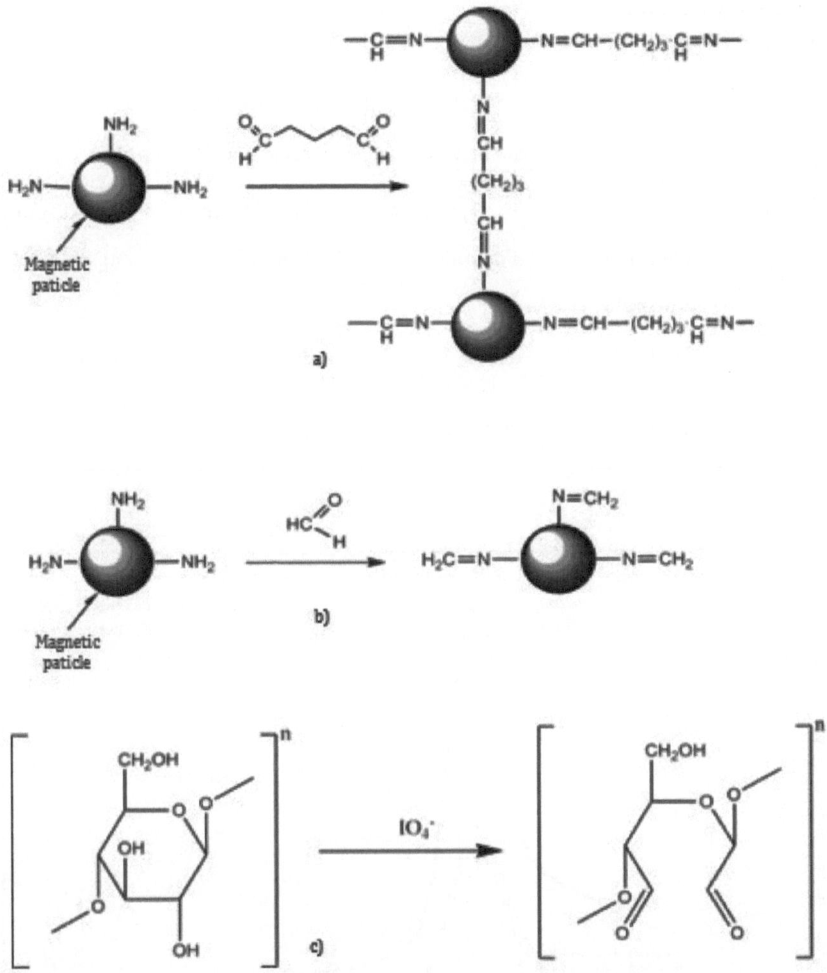

FIGURE 12.1 Modification of proteins (albumin or gelatin) NH_2-groups by glutaraldehyde (a) and formaldehydes (b) activation of dextran OH-groups with $NaJO_4$

The results on immobilization of hemoglobin on modified ferrocomposites particles are shown in and Table 12.2 and Figure 12.3.

TABLE 12.2 Sorption efficiency of modified ferrocomposites to hemoglobin (composite/Hb, w/w = 20) in physiological solution at pH 7.4

	Sorption, Average ± SD(%) Absorptive Capacity, Average ± SD (mg/g)			
Types of Composit	**Fe + Gelatin + Formald**	**Fe_3O_4 + Gelatin + Formald**	**Fe + Dextran + $NaJO_4$**	**Fe + Albumin+ Formald**
	38.0 ± 4.6	41.0 ± 4.2	94.5 ± 11.3	68.4 ± 7.5
	19.0 ± 2.3	20.5 ± 2.1	47.2 ± 5.6	34.2 ± 3.8

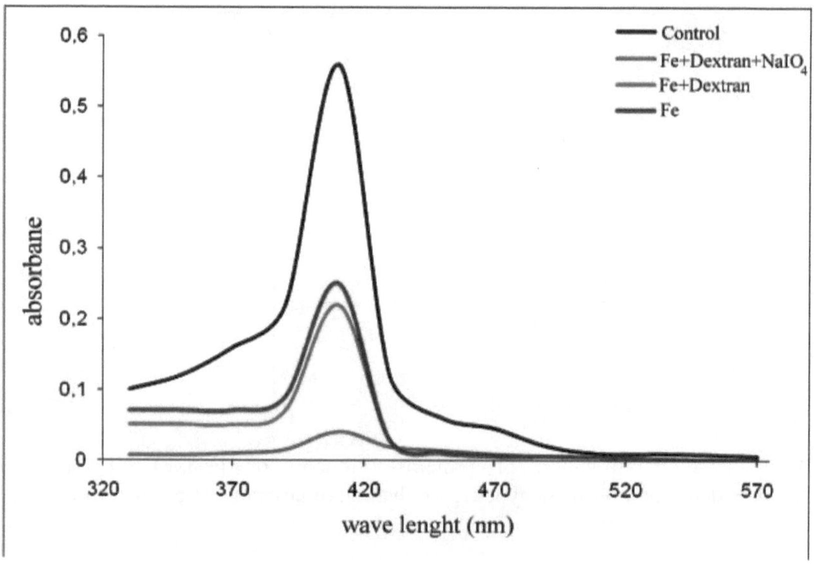

FIGURE 12.2 Absorption spectra of hemoglobin after immobilization on dextran-modified Fe-particles (Fe/Hb, w/w = 20) in physiological solution at pH 7.4.

Figure 12.2 and Table 12.2 demonstrates that iron particles covered by dextran and activated by $NaJO_4$ have shown maximal sorption efficiency to hemoglobin, which is, obviously, accounted for forming hydrogen connections between aldehydes groups of dextran and amino-groups of hemoglobin.

Sorption efficiency of modified ferrocomposites to hemoglobin in a model biological liquid (0.6% albumin in physiological solution) is presented in Table 12.3.

TABLE 12.3 Sorption efficiency of modified ferrocomposites to hemoglobin (composite/Hb, w/w = 10) in 0.6 percent albumin in physiological solution at pH 7.4

	Sorption, Average ± SD (%) Absorptive Capacity, Average ± SD (mg/g)			
Types of Composites	**Fe_3O_4 + Gelatin + Formald**	**Fe + Gelatin + Formald**	**Fe + Albumin + Formald**	**Fe + Dextran + $NaJO_4$**
	8.2 ± 1.7	16.3 ± 2.4	32.6 ± 3.8	25.0 ± 3.2
	8.2 ± 1.7	16.3 ± 2.4	32.6 ± 3.8	25.0 ± 3.2

Results on sorption efficiency of modified magnetite and iron particles to hemoglobin in model biological liquid (Table 12.3) showed, that maximal absorptive capacity is manifested with Fe-particles, modified by albumin (32.6 mg/g) and dextran (25.0 mg/g). These values are lower than those in physiological solution. This can be explained by that fact that the decreasing of the sorption process velocity is because of increasing of solution viscosity. In fact, the sorption efficiency increases with increasing of the incubation time from 30 to 60 sec. The interesting results on sorption efficiency of hemoglobin, carboxyhrmoglobin, and methemoglobin on gelatin-modified Fe-particles have been found in donor blood. These values are equal to 60.7 percent, 52.9 percent, and 22.5 percent respectively.

It is important to emphasize that adsorption of albumin on unmodified particles reached up to 40 percent for all composites types, and after modification of composites surface, adsorption of albumin was not more than 10 percent.

The sorption efficiency results of different chemical content of modified Ferro composites to phenobarbital in physiological solution are represented in Figures 12.3 and 12.4 and in Tables 12.4 and 12.5.

TABLE 12.4 Sorption efficiency of different chemical content of modified ferrocomposites to phenobarbital (PhB), (composite/PhB, w/w=20) in physiological solution at pH 7.4

	Sorption, Average ± SD (%) Absorptive Capacity, Average ± SD (mg/g)			
Types of Composites	**Fe Unmodified**	**Fe + Gel + Formald**	**Fe + Albumin + Formald**	**Fe-Al_2O_3**
	18.9 ± 1.8	36.2 ± 2.3	51.4 ± 5.8	14.2 ± 1.5
	9.4 ± 0.9	18.1 ± 1.2	25.7 ± 2.9	7.0 ± 0.7

Table 12.4 shows that modification of Fe-microparticles surface by albumin lead to considerable increase of phenobarbital immobilization: from 18.9 to 51.4 percent. The immobilization is realized, probably, by means of conjugation of phenobarbital with carboxilate- groups of albumin. In Figures 12.3 and 12.4 are shown absorption spectra of phenobarbital after immobilization with different chemical content microparticles.

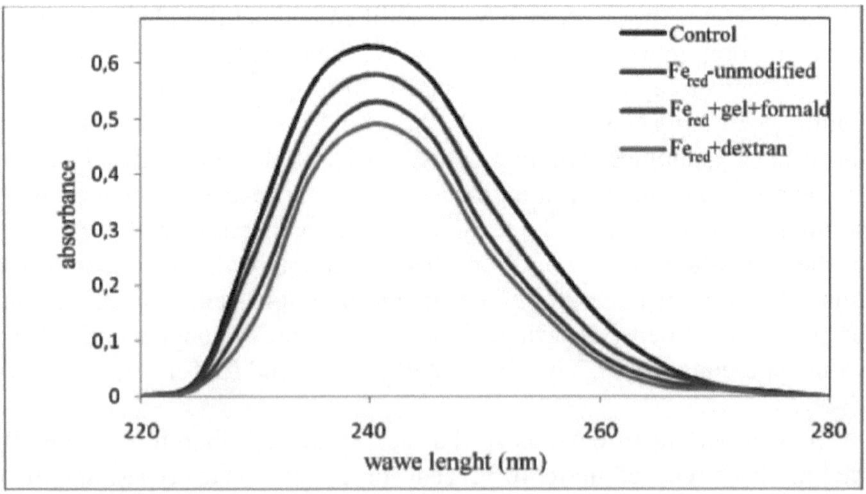

FIGURE 12.3 Absorption spectra of phenobarbital after immobilization on different chemical content of Fe-particles (Fe/PhB, w/w = 20) in physiological solution at pH 7.4.

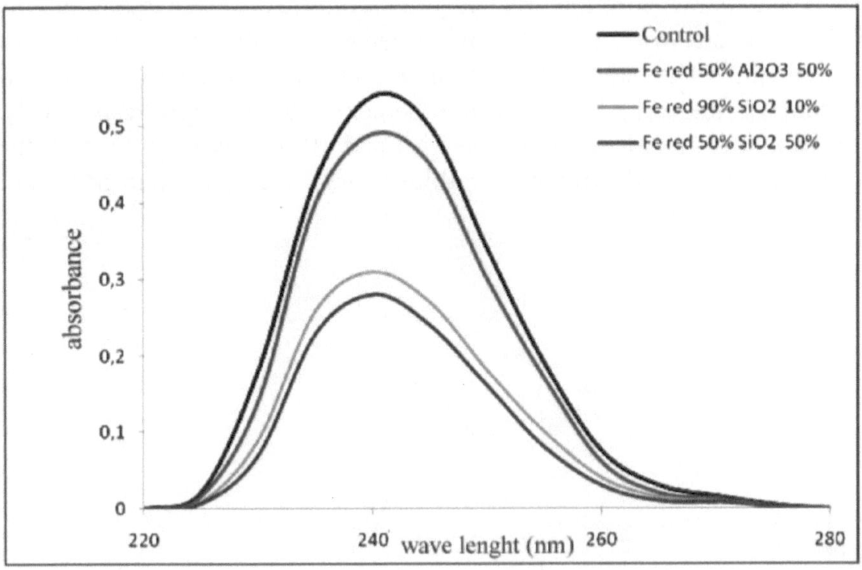

FIGURE 12.4 Absorption spectra of phenobarbital after immobilization on different chemical content of Fe-particles (Fe/PhB, w/w = 20) in physiological solution at pH 7.4.

TABLE 12.5 Sorption efficiency of different chemical content of ferrocomposites to phenobarbital (PhB), (composite/PhB, w/w = 20) in physiological solution at pH 7.4.

	Sorption, Average ± SD (%) Absorptive Capacity, Average ± SD (mg/g)			
Types of Composites	**Fe Unmodified**	**Fe + Dex**	**Fe-Silica (Fe90%, SiO210%)**	**Fe-Silica (Fe50%, $SiO_2$50%)**
	18.9 ± 1.8	15.9 ± 1.4	42.6 ± 4.2	48.1 ± 6.4
	9.4 ± 0.9	7.9 ± 0.7	21.3 ± 2.1	24.1 ± 3.2

Maximal values of sorption efficiency of phenobarbital have demonstrated on Fe-silica composites. Formation of hydrogen connections plays, apparently, a prevailing role in immobilization phenobarbital on Fe-silica composites. (Figure 12.4 and Table 12.5)

In Figure 12.5 are shown spectra of barbituric acid after immobilization on $FeCSiO_2$ microparticles at different weight ratios of composite/

BA. The maximal sorption efficiency of barbituric acid was found for FeC SiO_2 composite with a content of 50 percent Fe, 40 percent C, and 10 percent SiO_2. The values of sorption and absorptive capacity of barbituric acid for this composite were 75.0 percent and 58.0 mg/g respectively, and weight ratio $FeCSiO_2$/BA, were 50 and 10 respectively. Apparently, it shows physical adsorption of barbituric acid occurs in microporous of composite.

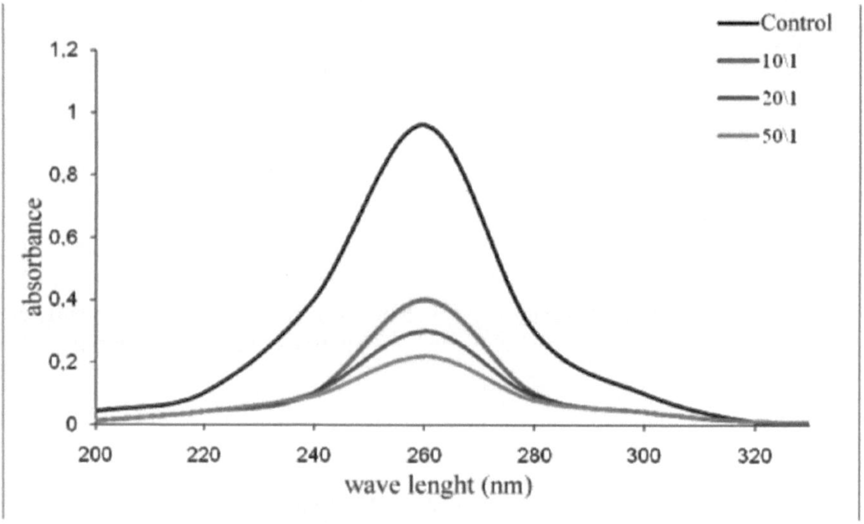

FIGURE 12.5 Absorption spectra of barbituric acid after immobilization on $FeCSiO_2$ particles at weight ratio composite/ BA were 10, 20, and 50.

12.4 IMMOBILIZATION AND DESORPTION OF L-BOROPHENILALANIN

In Figure 12.6 are shown spectra of L-BPA after immobilization on FeC microparticles at different weight ratios composite/L-BPA. Apparently immobilization occurs by physical adsorption into porous of composite. The highest absorption capacity of L-BPA for this composite 78.0 mg/g was detected at weight ratio composite/L-BPA as five. The maximal adsorption capacity of L-BPA 160.0 mg/g was reached for dextran-modified iron-particles.

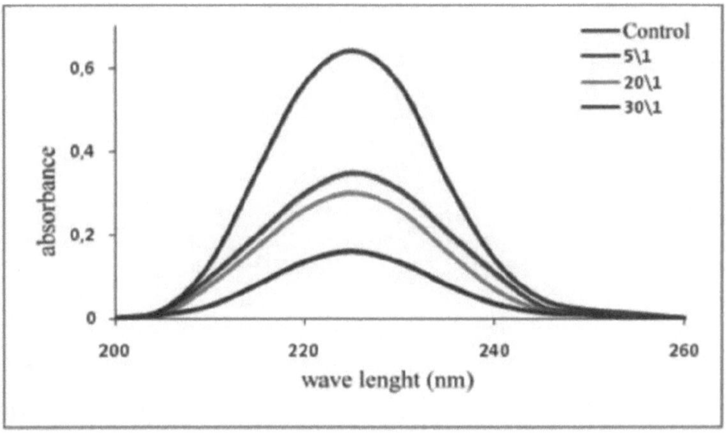

FIGURE 12.6 Absorption spectra of L-BPA after immobilization on FeC microparticles at different weight ratios composite/L-BPA.

The desorption of L-BPA at λ 225 nm from FeC composite and dextran-modified Fe-particles are presented in Figures 12.7 and 12.8. Analysis of the results on desorption has shown that quantity of the desorbed L-BPA from magnetic-operated preparations is enough to create therapeutic concentration of boron atoms in tumor. Nevertheless it is required to continue investigations in order to chose the optimum ferrocomposites types with more longer desorption time for working out magnetic-operated preparations of L-BPA on their basis.

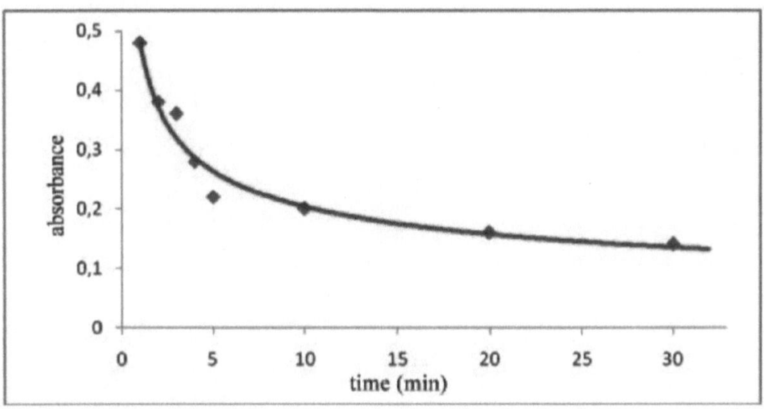

FIGURE 12.7 The dynamics of L-BPA desorption (λ225 nm) from FeC composite in 0.6 percent albumin (T 37°C, pH 7.4)

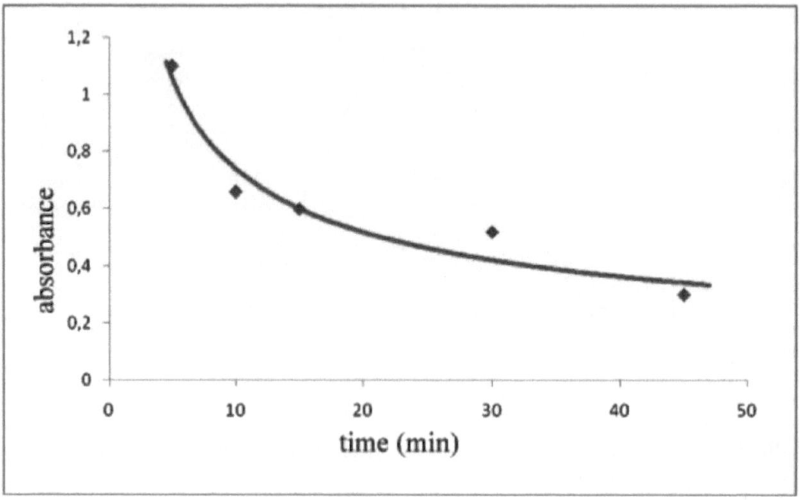

FIGURE 12.8 The dynamics of L-BPA desorption (λ225 nm) from dextran-modified Fe-particles in 0.6 percent albumin (T 37°C, pH 7.4).

Spectrophotometric study of the reaction interaction L-BPA with dextran in the water solutions showed its conjugation with dextran (Figure 12.9).

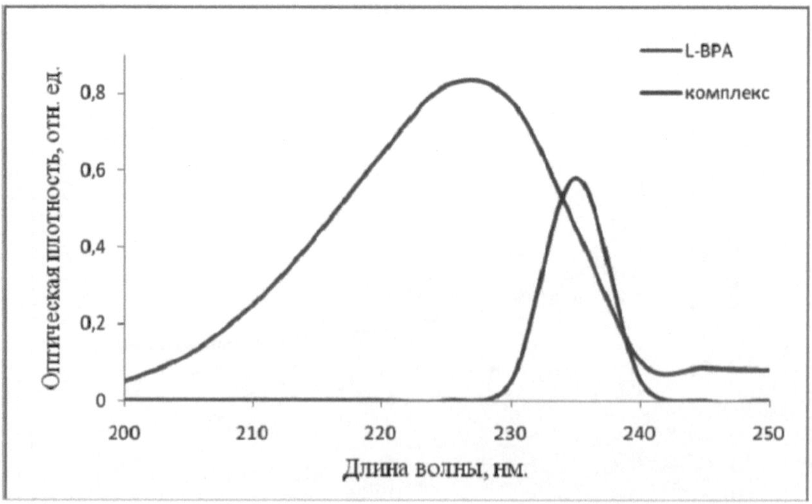

FIGURE 12.9 Conjugation of L-BPA with dextran: T 20°C, pH 7.0, incubation time 1 min.

12.5 SUMMARY

We have worked out new methods to modify the surface of the ferrocomposites microsized particles (iron, magnetite, iron-carbon) with albumin, gelatin or dextran and have been studied their sorption efficiency and the same for composite iron-silica and iron-carbon-silica to bovine hemoglobin and barbiturates (sodium phenobarbital and barbituric acid). Optimal Fe-composites for hemoglobin immobilization are with albumin and dextran modified microsized Fe-particles, for phenobarbital the albumin-modified Fe-particles and Fe-silica composite; for barbituric acid the FeC-silica composite. These ferrocomposites can be recommended for use as sorbents for extracorporal detoxification of patient's blood and purification of conserved donor blood from free hemoglobin and barbiturates by the method of magnetic separation. Dextran modified microsized Fe-particles are perspective as carriers for magnetically-guided targeted delivery of L-BPA at Boron Neutron Capture of Tumor Therapy.

KEYWORDS

- **Barbiturates**
- **Hemoglobin**
- **Immobilization of biologically active compounds**
- **L-borophenilalanin**
- **Microsized particles of ferrocomposites**
- **Modified surface**
- **Sorption efficiency**

REFERENCES

1. Komissarova, L. Kh.; Filippov, V. I.; and Kuznetsov, A. A.; et al. In: Mater of the 1st Symposium on Application of Biomagnetic Carriers in Medicine. Moscow; **2002,** 68–76 pp.
2. Brusentsov, N. A.; Bayburtskiy, F. S.; and Komissarova, L. Kh.; et al. Biocatalytic Technology and Nanotechnology. Moscow: Nova Science Publishers; **2004,** 59–66 pp.
3. Yanovsky, Y. G.; Komissarova, L. Kh.; and Danilin, A. N.; et al. *Solid State Phenomena.* **2009,** 152–153, 403–406.

4. Zhiwei Li; Chao Wang; and Liang Cheng; et al. *Biomaterials.* **2013,** *4,* 9160–9170.
5. Kutushov, M. W.; Komissarova, L. Kh.; and Gluchoedov, N. P.; Russian Patent No. 210952. **1998**.
6. Bunis, R. J.; Riley, K.J.; and Marling, O. K.; In: Research and Development in Neutron Capture Therapy. Ed. Monduzzi. Bologna. **2002,** 405 pp.

CHAPTER 13

A THEORY OF HEAVY ATOMS: A NEW RELATIVISTIC APPROACH IN MOMENTUM REPRESENTATION

B. K. NOVOSADOV

CONTENTS

13.1 AIM AND BACKGROUND

The development of high energy physics and chemistry leads to a necessity of seeking for and employing many-particle relativistic equations. A goal of this paper is to propose some relativistic models and to give methods of their solving for heavy atoms. A new relativistic approach in the theory of heavy atoms has been suggested in momentum representation. A scalar relativistic equation as an approximation to the equation system has been suggested, taking into account the spin-relativistic kinematics of atomic electrons.

13.2 INTRODUCTION

The problem studies on the electronic structure of heavy atoms face some theoretical difficulties in describing physical behavior of many-particle system in accordance with the relativity theory. Spectroscopy of heavy atoms has given a large amount of data that needs proper interpretation on the base of relativistic theoretical models. So far, there does not exists many-electron relativistic approaches for Coulombic systems, from which one could obtain approximations (even Hartree–Fock models) for the purpose of systematical study in the field of atomic and molecular spectroscopy and quantum chemistry. The development of high energy physics and chemistry leads to the necessity of seeking and employing many-particle relativistic equations. A goal of this paper is to suggest some relativistic models and to give methods of their solving for heavy atoms. Some efforts were be made for obtaining asymptotic properties of wave functions and spectra of many-electron stationary systems.

(i) Relativistic equation system in coordinate space of particles:

Classical energy for a system of free particles may be written as follows

$$E = \sum_{k=1}^{n} \sqrt{c^2 p_k^2 + m_k^2 c^4} \tag{13.1}$$

We may consider this expression like a root of some eigenvalue problem. For a system of non-interacting particles (electrons) this root is a sum of eigenvalues belonging to Dirac equations, or another one-particle relativistic equation system, for example, those in quaternionic representation. We use the both possibilities in our paper.

The main relation in relativistic physics connects energy and momentum taking into account the twofold degeneracy by spin

$$\left(\frac{E^2}{c^2} - p^2 - m^2c^2\right)^2 = 0\cdot \tag{13.2}$$

The one-particle Dirac system of four equations may be written as follows

$$\begin{pmatrix} c\sigma p & mc^2 \\ mc^2 & \text{-}c\sigma p \end{pmatrix}\begin{pmatrix} g(r) \\ u(r) \end{pmatrix} = E\begin{pmatrix} g(r) \\ u(r) \end{pmatrix}. \tag{13.3}$$

The kinematic matrix differs from the original Dirac one up to an orthogonal transformation of spinor components, but Eq. (13.3) is more convenient for deriving spin-relativistic members to non-relativistic expression for the particle energy (we do not face fractions with singular denominators in the variable r, when a Coulombic potential is included). Equation (13.3) has four roots, of which two are negative and has no physical sense, however the eigenvalue spectrum proves to be unlimited and there does not exist a lower limit to formulate a variational principle for the Dirac equation directly. This is one of the difficulties in numerical analysis of the relativistic equations for many-particle systems in quantum mechanics.

(ii) Fourier-transformation to momentum space of particles:

The particle coordinates and momentum are conjugate variables in quantum mechanics. A transfer to the momentum representation for the wave equation can be made by the Fourier-transformation of the wave function and operators. One has

$$\varphi(\mathbf{p}_1,\dots,\mathbf{p}_n) = (2\pi)^{-3n/2}\int \exp\left(-i\sum_{k=1}^{n}\mathbf{p}_k\mathbf{r}_k\right)\psi(\mathbf{r}_1,\dots,\mathbf{r}_n)\prod_{k=1}^{n} d^3\mathbf{r}_k \tag{13.4}$$

The Coulomb potential is transformed by the formula

$$\int e^{-i\mathbf{pr}} r^{-1} d^3\mathbf{r} = 4\pi p^{-2} \tag{13.5}$$

The inverse Fourier-transformation of the Coulomb potential is calculated as follows

$$2\pi^2 r^{-1} = \int e^{i\mathbf{pr}} p^{-2} d^3\mathbf{p} \tag{13.6}$$

The convolution theorem allows one to calculate the Fourier-transformation of the two functions product

$$\int \exp(-i\mathbf{pr}) f(\mathbf{r}) g(\mathbf{r}) d^3\mathbf{r} = \int \bar{f}(\mathbf{k}) \bar{g}(\mathbf{p}-\mathbf{k}) d^3\mathbf{k} \tag{13.7}$$

where

$$\bar{f}(\mathbf{k}) = (2\pi)^{-3/2} \int \exp(-i\mathbf{kr}) f(\mathbf{r}) d^3\mathbf{r} \tag{13.8}$$

A proof of the formulae given can be made by using the properties of the Dirac δ-function

$$\delta(\mathbf{r}) = (2\pi)^{-3} \int \exp(i\mathbf{pr}) d^3\mathbf{p}\,. \tag{13.9}$$

The Fourier-transformation of the Dirac equation converts the momentum into a *c*-number, while the product of the potential function and a bispinor turns into an integral in which the first one becomes the kernel of the equation integral operator. Write down the integral Dirac equation for hydrogen atom (the proton is considered fixed).

$$\begin{pmatrix} c\sigma p & mc^2 \\ mc^2 & -c\sigma p \end{pmatrix} \begin{pmatrix} \bar{g}(p) \\ \bar{u}(p) \end{pmatrix} = \left(E - V(p, p')\right) \begin{pmatrix} \bar{g}(p) \\ \bar{u}(p) \end{pmatrix}. \tag{13.10}$$

Here the product of the potential function and the bispinor ought to be understood like the integral expression

$$V(p, p') \begin{pmatrix} \bar{g}(p) \\ \bar{u}(p) \end{pmatrix} = \frac{1}{2\pi^2} \int \frac{-Ze^2}{(p - p')^2} \begin{pmatrix} \bar{g}(p') \\ \bar{u}(p') \end{pmatrix} d^3 p'. \tag{13.11}$$

An analogous Fourier-transformation allows one to write down in the momentum space many-particle relativistic equations, which are given below.

(iii) Eigenvectors of the Dirac kinematic matrix in the momentum space:

The relationship Eq. (13.2) can be considered as a *determinant* of an eigenvalue problem with the Dirac matrix B and a column-function $\varphi(t)$ with four components

$$\mathrm{B}\varphi = \frac{E}{c}\varphi, \tag{13.12}$$

where the matrix B is as follows

$$\mathbf{B} = \begin{bmatrix} m_0 c & 0 & p_z & p_- \\ 0 & m_0 c & p_+ & -p_z \\ p_z & p_- & -m_0 c & 0 \\ p_+ & -p_z & 0 & -m_0 c \end{bmatrix} \tag{13.13}$$

with the momentum cyclic components. One can easily verify that the decision problem condition for the homogeneous Eq. (13.12)

$$\det\left(\mathrm{B} - \frac{E}{c}\mathrm{I}_4\right) = 0, \tag{13.14}$$

where I_4 is the unit matrix of the order four, coincides with the Eq. (13.2). When the momentum components in the matrix given are *c*-numbers, then the column-function $\varphi(t)$ represents a set of four numbers, which are expressed via the B matrix elements.

Introduce a vector matrix (Clifford unit vector)

$$\sigma = \begin{pmatrix} n_z & n_x - in_y \\ n_x + in_y & -n_z \end{pmatrix}, \tag{13.15}$$

Where the matrix elements are cyclic unit vectors of the Cartesian coordinate system, $i = \sqrt{-1}$, and briefly $\mathbf{n}_- = \mathbf{n}_x - i\mathbf{n}_y$, $\mathbf{n}_+ = \mathbf{n}_x + i\mathbf{n}_y$, then the matrix (13.15) may be written concisely as

$$\sigma = \begin{pmatrix} n_z & n_- \\ n_+ & -n_z \end{pmatrix}, \tag{13.16}$$

or on the basis of Pauli matrices

$$\sigma = \sigma_x n_x + \sigma_y n_y + \sigma_z n_z \,. \tag{13.17}$$

The matrix B may be written like a block matrix of the order 2

$$B = \begin{pmatrix} m_0 c I_2 & \sigma p \\ \sigma p & -m_0 c I_2 \end{pmatrix}, \tag{13.18}$$

where $I_2 = \begin{pmatrix} 1 & 0 \\ 0 & 1 \end{pmatrix}$ and the momentum are given in the Clifford algebra.

To diagonalize the matrix B one may notice that the blocks along the main diagonal are proportional to the unit matrix of the order two, therefore they are invariable. If one diagonalizes first the momentum blocks, which are Hermite matrices,

$$\Pi = \sigma p = \begin{bmatrix} p_z & p_- \\ p_+ & -p_z \end{bmatrix}. \tag{13.19}$$

the matrix Π may be written down via the eigenvalue matrix

$$\Lambda = \begin{bmatrix} \lambda_1 & 0 \\ 0 & \lambda_2 \end{bmatrix}, \tag{13.20}$$

where $\lambda_1 = p$, $\lambda_2 = -p$, $p = \sqrt{p_x^2 + p_y^2 + p_z^2}$, are spectral decomposition

$$\Pi = C_1^+ \Lambda C_1, \tag{13.21}$$

and the cross indicates the Hermite conjugation of the eigenvector unitary matrix C_1. Solving the matrix equation $\Pi \mathbf{c} = \lambda \mathbf{c}$, one obtains the eigenvector matrix sought for

$$\mathbf{C}_1 = \begin{bmatrix} \sqrt{\dfrac{p+p_z}{2p}} & \dfrac{-p_-}{\sqrt{2p(p+p_z)}} \\ \dfrac{p_+}{\sqrt{2p(p+p_z)}} & \sqrt{\dfrac{p+p_z}{2p}} \end{bmatrix}. \tag{13.22}$$

Transformation

$$U_1 = \begin{bmatrix} C_1 & 0 \\ 0 & C_1 \end{bmatrix} \tag{13.23}$$

brings the matrix Π, as has been said earlier, into the diagonal form without changing the diagonal blocks in Eq. (13.18). As a result the matrix B has been transformed to a simpler form with a more number of zero elements

$$U_1^+ BU_1 = \begin{bmatrix} m_0c & 0 & p & 0 \\ 0 & m_0c & 0 & -p \\ p & 0 & -m_0c & 0 \\ 0 & -p & 0 & -m_0c \end{bmatrix}. \qquad (13.24)$$

Besides the zeros are arranged in the chess order, so the permutation of the second and third rows and columns of the matrix given by the matrix

$$P_{23} = \begin{bmatrix} 1 & 0 & 0 & 0 \\ 0 & 0 & 1 & 0 \\ 0 & 1 & 0 & 0 \\ 0 & 0 & 0 & 1 \end{bmatrix}, \qquad (13.25)$$

brings the matrix B into a block-diagonal form

$$P_{23}U_1^+ BU_1P_{23} = \begin{bmatrix} m_0c & p & 0 & 0 \\ p & -m_0c & 0 & 0 \\ 0 & 0 & m_0c & -p \\ 0 & 0 & -p & -m_0c \end{bmatrix}, \qquad (13.26)$$

with the blocks of like structures. Usage of the orthogonal transformation

$$\mathrm{F} = \begin{pmatrix} \mathrm{I}_2 & 0 \\ 0 & \mathrm{F}_2 \end{pmatrix}, \qquad (13.27a)$$

where the matrix I_2 is the diagonal unit matrix of the order two, F_2 is the diagonal of the form

$$\mathrm{F}_2 = \begin{pmatrix} 1 & 0 \\ 0 & -1 \end{pmatrix}, \qquad (13.27b)$$

brings the matrix (13.26) into the matrix with identical blocks, which are diagonalized by the same orthogonal matrix C_2 of a general form

$$\mathrm{C}_2 = \begin{bmatrix} \cos\varphi & -\sin\varphi \\ \sin\varphi & \cos\varphi \end{bmatrix}. \qquad (13.28)$$

The eigenvalues of the block matrices (13.26) are as follows

$$\frac{E_1}{c} = \sqrt{m_0^2 c^2 + p^2}\ , \tag{13.29}$$

For the angle φ a relationship $tg2\varphi = \frac{p}{m_0 c}$ takes place, from which one finds, using the known relationship $tg2\varphi = \frac{2t}{1+t^2}$, the parameter $t = tg\varphi$

$$t = \frac{p}{m_0 c + \sqrt{m_0^2 c^2 + p^2}}\ . \tag{13.30}$$

Thus, the elements of the orthogonal matrix (13.28) are calculated as follows

$$\cos\varphi = \frac{1}{\sqrt{1+t^2}}\ ,\quad \sin\varphi = t\cos\varphi \qquad (13.31$$

As a result, the matrix B is brought into the diagonal form with the eigenvalues Eq. (13.29) with the help of the set of four matrix transformations

$$\tilde{U}_2 F P_{23} U_1^{+} B U_1 P_{23} F U_2 = \frac{1}{c}\begin{bmatrix} E_1 & 0 & 0 & 0 \\ 0 & E_2 & 0 & 0 \\ 0 & 0 & E_1 & 0 \\ 0 & 0 & 0 & E_2 \end{bmatrix}, \tag{13.32}$$

where the orthogonal matrix U_2 is as follows

$$\mathbf{U_2} = \begin{bmatrix} \mathbf{C_2} & \mathbf{0} \\ \mathbf{0} & \mathbf{C_2} \end{bmatrix}. \tag{13.33}$$

Considering the first column of the eigenvector matrix for the matrix B, one gets normalized 1 eigenspinor

$$\varphi_1 = \begin{pmatrix} \sqrt{\dfrac{p+p_z}{2p}} \dfrac{m_0c+\sqrt{m_0^2c^2+p^2}}{\sqrt{p^2+(m_0c+\sqrt{m_0^2c^2+p^2})^2}} \\ \dfrac{p_+}{\sqrt{2p(p+p_z)}} \dfrac{m_0c+\sqrt{m_0^2c^2+p^2}}{\sqrt{p^2+(m_0c+\sqrt{m_0^2c^2+p^2})^2}} \\ \sqrt{\dfrac{p+p_z}{2p}} \dfrac{p}{\sqrt{p^2+(m_0c+\sqrt{m_0^2c^2+p^2})^2}} \\ \dfrac{p_+}{\sqrt{2p(p+p_z)}} \dfrac{p}{\sqrt{p^2+(m_0c+\sqrt{m_0^2c^2+p^2})^2}} \end{pmatrix}. \quad (13.34)$$

Denoting the matrix elements in Eq. (13.22) by c_{11}, c_{12}. c_{21}, c_{22}, sine and cosine in the matrix (13.28) by s and c, we can write down an explicit form of the normalized eigenvectors matrix for the matrix B

$$\mathrm{U_1P_{23}FU_2} = \begin{bmatrix} \tilde{n}_{11}\tilde{n} & -\tilde{n}_{11}s & c_{12}c & -c_{12}s \\ c_{21}c & -c_{21}s & c_{22}c & -c_{22}s \\ c_{11}s & c_{11}c & -c_{12}s & -c_{12}c \\ c_{21}s & c_{21}c & -c_{22}s & -c_{22}c \end{bmatrix}. \quad (13.35)$$

Here, the letter c cannot be confused with the light velocity in the matrix B (13.13).

In the nonrelativistic limit, when the momentum becomes much less than the quantity m_0c, the "positron" components of the column $\ddot{o}_1$ tend to become zero. In this case, the parameter Eq. (13.20) is equal to zero, and the one obtained in accordance with Eq. (13.31) and that in the matrix (13.35), the c =1 and s = 0. The elements c_{ij} do not depend on the radial momentum and the matrix (13.22) may be expressed via the angle variables θ, φ, which define a direction of the particle momentum vector

$$C_1 = \begin{pmatrix} \sqrt{\dfrac{1+\cos\theta}{2}} & -e^{-i\varphi}\sqrt{\dfrac{1-\cos\theta}{2}} \\ e^{i\varphi}\sqrt{\dfrac{1-\cos\theta}{2}} & \sqrt{\dfrac{1+\cos\theta}{2}} \end{pmatrix} = \begin{pmatrix} \cos\left(\dfrac{\theta}{2}\right) & -\sin\left(\dfrac{\theta}{2}\right)e^{-i\varphi} \\ \sin\left(\dfrac{\theta}{2}\right)e^{i\varphi} & \cos\left(\dfrac{\theta}{2}\right) \end{pmatrix} \quad (13.36)$$

Thus, the nonrelativistic bispinor is of the form

$$\varphi_1 = \begin{pmatrix} c_{11} \\ c_{21} \\ 0 \\ 0 \end{pmatrix}. \tag{13.37}$$

Analogously we obtain another columns for the nonrelativistic bispinors of which only one has a physical meaning and correspond to the positive eigenvalue of the matrix B. The eigenvectors Eq. (13.35) will be used below.

(iv) Reduction of the bispinor Dirac equation to an integral form:

A solution of the integral Dirac equation is given in [1]. Here we suggest a general method of transformation the relativistic equations to a convenient integral form for their analysis. Making use of the theorem from the matrix theory, Hermitian matrix A can be written down like the spectral resolution over eigenvectors $\mathbf{c}_k$ as follows

$$\text{A}=\sum_{k=1}^{n} c_k^* \lambda_k \tilde{c}_k\,. \tag{13.38}$$

In this formula the wave line denotes the row-vector. A number between the vectors is their eigenvalue of the matrix. In an analogous manner one can write down the kinematic matrix in the left hand side of the Eq. (13.43), taking into consideration that it is the c-number matrix,

$$\sum_{k=1}^{4} c_k^* \lambda_k \tilde{c}_k \begin{pmatrix} \overline{g}(p) \\ \overline{u}(p) \end{pmatrix} = \left(E - V(p, p')\right) \begin{pmatrix} \overline{g}(p) \\ \overline{u}(p) \end{pmatrix}. \tag{13.39}$$

The product of a vector row and a bispinor column is a scalar function of the momentum

$$\tilde{\mathbf{c}}_k \begin{pmatrix} \overline{g}(\mathbf{p}) \\ \overline{u}(\mathbf{p}) \end{pmatrix} = \varphi_k(\mathbf{p}). \tag{13.40}$$

The eigenvectors $\mathbf{c}_k$ are given by the formulae (13.34) and (13.35). The eigenvalues of the Dirac equation matrix are equal to (where m is the electron rest mass)

$$\lambda_{1,3} = \sqrt{m^2 c^4 + c^2 p^2}, \quad \lambda_{2,4} = -\sqrt{m^2 c^4 + c^2 p^2}. \tag{13.41}$$

Multiplying the Eq. (13.39) to the left by the vector row $\tilde{\mathbf{c}}_1$, and taking into account the notation Eq. (13.40) and the orthogonality of the Dirac matrix eigenvectors, we arrive at a scalar equation relative to the function $\varphi_1(\mathbf{p})$

$$\left(\sqrt{m^2c^4+c^2p^2}-E\right)\varphi_1(\mathbf{p})=\frac{Ze^2}{2\pi^2}\int\frac{1}{(\mathbf{p}-\mathbf{p}')^2}\tilde{\mathbf{c}}_1(\mathbf{p})\psi(\mathbf{p}')d^3\mathbf{p}'. \quad (13.42)$$

The scalar product of the vector-row and vector-column $\tilde{c}_1(p)\psi(p')$ under the integral is a scalar function depending on the two vector arguments $\mathbf{p}$ and $\mathbf{p}'$, therefore one ought to transform this product to the scalar functions $\varphi_k(\mathbf{p}')$. Note that the unit matrix of the order four can be represented like the decomposition over eigenvectors $\mathbf{c}_k$ of the matrix (13.35). One has got

$$\mathrm{I}_4=\sum_{k=1}^{4}\mathbf{c}_k^*(\mathbf{p}')\tilde{\mathbf{c}}_k(\mathbf{p}'). \quad (13.43)$$

Substituting this matrix into the integral of Eq. (13.42), we arrive at the integral equation as follows

$$\left(\sqrt{m^2c^4+c^2p^2}-E\right)\varphi_1(\mathbf{p})=\frac{Ze^2}{2\pi^2}\int\frac{\tilde{\mathbf{c}}_1(\mathbf{p})}{(\mathbf{p}-\mathbf{p}')^2}\sum_{k=1}^{4}\mathbf{c}_k^*(\mathbf{p}')\varphi_k(\mathbf{p}')d^3\mathbf{p}'. \quad (13.44)$$

In the same way the equations for the functions $\varphi_k(\mathbf{p})$, $k=2,\ 3,\ 4$, can be obtained, with the difference that the radicals of the second and fourth equations will be taken with the minus sign. The kernels of the integral equations obtained are scalar functions, because the product of vector-row and vector-column is the vector scalar product. As we see that these scalar multipliers are factorized by the variables $\mathbf{p}$ and $\mathbf{p}'$, hence subsequent solving of the relativistic equations system Eq. (13.42) can be made by the factorization of the Coulombic part of the integral operator kernel. This problem can be solved with the help of Fock resolution of that function over four-fold spherical harmonics [2], which used while solving the Schrödinger equation for the hydrogen atom.

The nonrelativistic approximation takes place, provided there is low electron momentum as compared with *mc*, and where the eigenvector $\mathbf{c}_k$

becomes unit and the scalar function $\varphi_1(\mathbf{p})$ remains only in the integral Eq. (13.44). Representing the radical as a series in p/mc and restriction of the latter two first terms, with denoting $\varepsilon = E - mc^2$, one arrives at the Schrödinger integral equation for an electron in the hydrogen atom

$$\left(\frac{p^2}{2m} - \varepsilon\right)\varphi_1(\mathbf{p}) = \frac{Ze^2}{2\pi^2}\int \frac{\varphi_1(\mathbf{p}')}{(\mathbf{p}-\mathbf{p}')^2} d^3\mathbf{p}'. \tag{13.45}$$

Solving this equation allows one to make evident the O(4) symmetry of the Coulombic problem in wave mechanics of the hydrogen atom, established in the classical Kepler problem [4, 5]. In the next section a solution of that integral equation will be obtained with the help of four-fold spherical harmonics.

(v) Solving the Schrödinger integral equation for the hydrogen atom:

We seek a solution of Eq. (13.9) by the Fock method [2], so that this problem for each electronic state becomes equivalent to that for the four-dimensional quantum rotator:

$$\left(p_0^2 + p^2\right)\ \psi(\mathbf{p}) = \pi^{-2}\int |\,\mathbf{p}-\mathbf{p}'|^{-2}\,\psi(\mathbf{p}')\,d^3\mathbf{p}' \tag{13.46}$$

where p_0 is the mean quadratic momentum, and $p_0^2 = -2\varepsilon > 0$ for bound states. The last condition means that the form $p_0^2 + p^2$ has the elliptic kind.

Introducing a four-dimensional momentum $p_4^2 = p_0^2 + p^2$ we define an angle variable α, then

$$\cos\alpha = \left(p_0^2 - p^2\right)/\left(p_0^2 + p^2\right),\ \sin\alpha = 2p_0 p/\left(p_0^2 + p^2\right), \tag{13.47}$$

where $\alpha \in [0,\ \pi]$. The relations (13.47) are the stereographic projection of the momentum p. The angle α together with the usual spherical angles $\theta,\ \varphi$ of the momentum $\mathbf{p}$ define the surface of the unit sphere in R^4. The distance between two points of the sphere is given by the arc length of the great circle, which goes through those points. For the unit sphere

R^4 this distance is equal to the central angle ω (in radians) between the radius-vectors of the points. So, $\cos\omega$ is as follows

$$\cos\omega = \cos\alpha\cos\alpha' + \sin\alpha\sin\alpha'\cos\gamma, \tag{13.48}$$

where

$$\cos\gamma = \cos\theta\cos\theta' + \sin\theta\sin\theta'\cos(\varphi - \varphi'). \tag{13.49}$$

Expressing the distance square between points $\mathbf{p}$ and $\mathbf{p}'$ in terms of $\cos\omega$, and making use of the relation (13.47), we get

$$(p - p')^2 = p^2 + (p')^2 - 2pp'\cos\gamma = p^2 + (p')^2 -$$

$$-(2p_0^2)^{-1}(p_0^2 + p^2)\left(p_0^2 + (p')^2\right)\sin\alpha\sin\alpha'\cos\gamma = \tag{13.50}$$

$$= (4p_0^2)^{-1}(p_0^2 + p^2)\left(p_0^2 + (p')^2\right)\left[\frac{4p_0^2\left(p^2 + (p')^2\right)}{(p_0^2 + p^2)\left(p_0^2 + (p')^2\right)} - 2\sin\alpha\sin\alpha'\cos\gamma\right]$$

A direct test gives

$$2 - 2\cos\alpha\,\cos\alpha' = \frac{4p_0^2\left(p^2 + (p')^2\right)}{(p_0^2 + p^2)\left(p_0^2 + (p')^2\right)}. \tag{13.50a}$$

With the help of the relationships given, we obtain the formula sought out

$$(\mathbf{p} - \mathbf{p}')^2 = (2p_0)^{-2}(p_0^2 + p^2)\left(p_0^2 + (p')^2\right)(2 - 2\cos\omega) \tag{13.51}$$

Write down the volume element $d^3\mathbf{p}$ in spherical coordinates

$$d^3\mathbf{p} = p^2 dp\,\sin\theta d\theta\,d\varphi\cdot \tag{13.52}$$

Using the relationships (13.47) we obtain

$$\cos(\alpha/2) = p_0\left(p_0^2 + p^2\right)^{-1/2}, \sin(\alpha/2) = p\left(p_0^2 + p^2\right)^{-1/2} \quad ,(13.53)$$

then the radial momentum can be expressed via the angle α

$$p = p_0 \ \mathrm{tg}(\alpha/2), \quad (13.54)$$

from which the differential dp can be easily calculated

$$dp = p_0\left[2 \ \cos^2(\alpha/2)\right]^{-1} d\alpha. \quad (13.55)$$

The volume element acquires the form in hyperspherical coordinates as follows

$$d^3\mathbf{p} = (2p_0)^{-3}\left(p_0^2 + p^2\right)^3 \sin^2\alpha \ \sin\theta \ d\alpha d\theta \ d\varphi \cdot \quad (13.56)$$

Denote the hypersurface element on the four-dimensional sphere

$$d\Omega_4 = \sin^2\alpha \ \sin\theta \ d\alpha d\theta \ d\varphi, \quad (13.57)$$

then the relationship (13.56) takes the form

$$d^3\mathbf{p} = (2p_0)^{-3}\left(p_0^2 + p^2\right)^3 d\Omega_4 . \quad (13.58)$$

For the wave function one obtains

$$\psi(\mathbf{p}) = a\left(p_0^2 + p^2\right)^{-2} \Psi(\alpha, \theta, \varphi), \quad (13.59)$$

where the coefficient $a = 2^{3/2}\pi^{-1}p_0^{5/2}$. With the help of the formulae (13.51), (13.58), and (13.59) the Schrödinger Eq. (13.46) is transformed to the Fock integral equation for the hydrogen atom

$$p_0\Psi(\alpha,\theta,\varphi) = \left(2\pi^2\right)^{-1} \int \left[4\sin^2(\omega/2)\right]^{-1} \Psi(\alpha',\theta',\varphi) \, d\Omega_4' \cdot \quad (13.60)$$

On the unit of hypersphere, the square of the distance between two points is

$$(\Delta\tilde{\mathbf{n}})^2 = (\tilde{\mathbf{n}}_1 - \tilde{\mathbf{n}}_2)^2, \tag{13.61}$$

where $\tilde{\mathbf{n}}_1, \tilde{\mathbf{n}}_2$ are unit vectors from the sphere center to its surface points. The Cartesian coordinates of a point on the 4-sphere can be expressed via the angle variables according to the formulae

$$t = \sin\alpha\ \sin\theta\ \sin\varphi,\ u = \sin\alpha\ \sin\theta\ \cos\varphi\ ,$$

$$v = \sin\alpha\ \cos\theta,\ w = \cos\alpha\ . \tag{13.62}$$

Taking into account that

$$t^2 + u^2 + v^2 + w^2 = \rho^2 = 1, \tag{13.63}$$

one obtains

$$(\Delta\tilde{\mathbf{n}})^2 = (\tilde{\mathbf{n}}_1 - \tilde{\mathbf{n}}_2)^2 = 2 - 2\cos\ \omega = 4\sin^2(\omega/2), \tag{13.64}$$

that coincides with the denominator of the function under the integral in Eq. (13.60)

Writing down the Laplace equation in R^4,

$$\left(\frac{\partial^2}{\partial t^2} + \frac{\partial^2}{\partial u^2} + \frac{\partial^2}{\partial v^2} + \frac{\partial^2}{\partial w^2}\right)\Psi(t,\ u,\ v,\ w) = 0. \tag{13.65}$$

We see this equation and the integral Eq. (13.60) are equivalent on the unit sphere surface. Solutions of these equations are the hyperspherical harmonics and are [2, 3] as follows

$$\Psi_{nlm}(\alpha,\ \theta,\ \varphi) = \Pi_n^l(\alpha)\ Y_{lm}(\theta,\ \varphi), \tag{13.66}$$

where

$$\Psi_{nlm}(\alpha,\ \theta,\ \varphi) = \Pi_n^l(\alpha)\ Y_{lm}(\theta,\ \varphi) \tag{13.67}$$

is a normalized to one spherical harmonic, provided $P_l^m(\theta)$ being a Legendre associated polynomial, and $\Pi_n^l(\alpha)$ being a Gegenbauer associated polynomial, which is connected with the Gegenbauer polynomial by a relationship

$$\Pi_n^l(\alpha) = b_{nl} \sin^l \alpha C_{n-l-1}^{l+1}(\cos\alpha), \quad (13.68)$$

where b_{lm}, b_{nl} are the normalization coefficients:

$$b_{lm} = (-1)^{(m+|m|)/2} \left[\frac{2l+1}{2} \frac{(l-|m|)\,!}{(l+|m|)\,!} \right]^{1/2},$$

$$b_{nl} = (-1)^{n+1} i^l 2\pi^{1/2} 2^l l! \left[n(n-l-1)\,!/(n+l)\,! \right]^{1/2}. \quad (13.69)$$

An explicit form of the function $\Psi_{nlm}(\alpha, \theta, \varphi)$ can be found if one expands the function $\left[4 sin^2(\omega/2)\right]^{-1}$ into a series over the Gegenbauer polynomials. Indeed, a series takes place [4, 5] representing a generalization of the Legendre series for the generating function

$$(\mathbf{\tilde{n}}_1 - \mathbf{\tilde{n}}_2)^{-2\lambda} = (\rho_1^2 + \rho_2^2 - 2\rho_1 \rho_2 \cos\omega)^{-\lambda} =$$

$$= \sum_{n=1}^{\infty} \frac{\left(\rho_1^2 + \rho_2^2 - |\rho_1^2 - \rho_2^2|\right)^{n+\lambda-1}}{2^{n+\lambda-1}(\rho_1 \rho_2)^{n+2\lambda-1}} C_{n-1}^{\lambda}(\cos\omega), \quad (13.70)$$

where $C_{n-1}^{\lambda}(\cos\omega)$ is a Gegenbauer polynomial, λ is a real number.

Taking into account the addition theorem [3] for the Gegenbauer polynomials

$$C_q^p(\cos\alpha\cos\alpha' + \sin\alpha\sin\alpha'\cos\gamma) =$$

$$= \frac{\Gamma(2p-1)}{[\Gamma(p)]^2} \sum_{l=0}^{q} \frac{2^{2l}\Gamma^2(p+l)\,(q-l)\,!\,(2l+2p-1)}{\Gamma(q+l+2p)} \times$$

$$\times \sin^l \alpha C_{q-l}^{p+l}(\cos\alpha) \sin^l \alpha' C_{q-l}^{p+l}(\cos\alpha') C_l^{p-1/2}(\cos\gamma) \quad (13.71)$$

we see it allows one to obtain a bilinear expansion of the kernel $\left[4\sin^2(\omega/2)\right]^{-1}$ over the hyperspherical harmonics.

$$\left[4\sin^2(\omega/2)\right]^{-1} = \sum_{n=1}^{\infty} \sum_{l=0}^{n-1} \sum_{m=-l}^{l} n^{-1}\Psi_{nlm}(\Omega_4)\,\Psi^*_{nlm}(\Omega'_4). \quad (13.72$$

Substituting this expansion into the Fock equation, then multiplying both parts of the equation by a complex conjugated hyperspherical function, and integrating over four-sphere surface, provided the hyperspherical harmonics orthonormality,

$$\int_0^{\pi}\int_0^{\pi}\int_0^{2\pi} \Psi_{nlm}(\Omega_4)\,\Psi^*_{n'l'm'}(\Omega_4)\,d\Omega_4 = 2\pi^2\delta_{nn'}\delta_{ll'}\delta_{mm'}, \quad (13.73)$$

where $\delta_{nn'}$ is the Kronecker symbol, we obtain for the parameter p_0 the following value

$$p_0 = n^{-1}. \quad (13.74)$$

The hydrogen atomic spectrum is calculated by the formula

$$\varepsilon = -\frac{1}{2}p_0^2. \quad (13.75)$$

Thus, the problem on the discrete spectrum of the hydrogen atom has been solved completely.

The hydrogen atomic continuous spectrum is characterized by the positive parameter ε to which the harmonics on the both surfaces of a two-sheeted hyperboloid correspond.

(vi) A remark on solving the relativistic integral equation system in momentum space:

Considering the relativistic equation system for a hydrogen-like atom,

$$\left(\sqrt{m^2c^4 + c^2p^2} - E\right)\varphi_i(\mathbf{p}) = \frac{Ze^2}{2\pi^2}\int \frac{\tilde{\mathbf{c}}_i(\mathbf{p})}{(\mathbf{p}-\mathbf{p}')^2}\sum_{k=1}^{4}\mathbf{c}_k^*(\mathbf{p}')\varphi_k(\mathbf{p}')d^3\mathbf{p}', \ (i = 1, \dots, 4) \quad (13.76)$$

and taking into account the factorization of the function $\tilde{c}_i(\mathbf{p})c_k^*(\mathbf{p}')$ by the variables, one can reduce the problem to solving the integral equation with a degenerate kernel and thus to a system of linear algebraic equations with a Hermite matrix of coefficients. A diagonalization of this matrix will come up with a solution of the initial equation, because coefficients in a series over the basic functions of the kernel bilinear decomposition will be found. It is clear that due to the spherical symmetry of the problem, the coefficient matrix in the equation system will acquire a block-diagonal form with finite block orders.

In addition, one can consider an approximate scalar equation for the first component φ_1 that corresponds to the spinor function with the predominant first component of the bispinor for the positive energy value

$$\left(\sqrt{m^2c^4+c^2p^2}-E\right)\varphi_1(\mathbf{p})=\frac{Ze^2}{2\pi^2}\int\frac{\tilde{\mathbf{c}}_1(\mathbf{p})\mathbf{c}_1^*(\mathbf{p}')}{(\mathbf{p}-\mathbf{p}')^2}\varphi_1(\mathbf{p}')d^3\mathbf{p}'. \tag{13.77}$$

This equation is an augmented one with respect to zero-spin integral equation when the bilinear kernel function becomes constant. In the given equation a modification of the Coulombic potential takes place in accordance with the relativistic kinematics by multiplying it into the bilinear by momenta function, which is formed from the kinematic matrix eigenvectors

$$\tilde{\mathbf{c}}_1(\mathbf{p})\mathbf{c}_1^*(\mathbf{p}')=c_{11}(\mathbf{p})c_{11}^*(\mathbf{p}')+c_{12}(\mathbf{p})c_{12}^*(\mathbf{p}')+c_{13}(\mathbf{p})c_{13}^*(\mathbf{p}')+c_{14}(\mathbf{p})c_{14}^*(\mathbf{p}'). \tag{13.78}$$

The bilinear form components are given by the formula (13.34). One can see that while neglecting the "positron" components c_{13} and c_{14}, we obtain the relativistic integral equation of the first order in energy with the spinor constituent unlike the second order as in energy Klein–Gordon equation

(vii) Relativistic equations for a many-electron system in the momentum space:

Some advantages of integral equations for a system of relativistic electrons are perceived while solving problems of electron scattering on atomic systems, the more so in experiments where electron momenta are measured. At the same time, as we have verified while researching the Dirac equation, in the momentum space one can easily pass to the nonrelativistic model of particle mechanics and obtain the relativistic corrections in an explicit form.

Writing down the system of relativistic equations for an atom

$$\sum_{a=1}^{n}\mathrm{I}_4\times\ldots\times\mathrm{B}_a^q\times\ldots\times\mathrm{I}_4\cdot\Psi+\left(\sum_{a=1}^{n}V_a+\sum_{a>b=1}^{n}V_{ab}\right)\mathrm{I}\cdot\Psi=E\Psi, \tag{13.79}$$

$$B_a^q=\begin{pmatrix} c\sigma p_a^q & mc^2I_2 \\ mc^2I_2 & -c\sigma p_a^q \end{pmatrix}, \tag{13.80}$$

where the index q points to the momentum being the differential operator.

Multiplying this system to the left by the exponential

$$\exp\left(-\sum_{k=1}^{n} i\mathbf{p}_k\mathbf{r}_k\right), \tag{13.81}$$

with the momentum being a c-number, and integrating the equations over electron coordinates, we arrive at the integral equation system

$$\sum_{a=1}^{n} \mathrm{I}_4 \times \ldots \times \mathrm{B}_a \times \ldots \times \mathrm{I}_4 \cdot \Phi(\mathbf{p}) +$$
$$+\frac{1}{(2\pi)^{3n/2}} \int \exp\left(\sum_{k=1}^{n} -i\mathbf{p}_k\mathbf{r}_k\right)\left(\sum_{a=1}^{n} V_a + \sum_{a>b=1}^{n} V_{ab}\right)\mathrm{I}\cdot\Psi(\mathbf{r})\, d^{3n}\mathbf{r} = E\Phi(\mathbf{p}). \tag{13.82}$$

Here the kinematic matrix in the direct product consists of the c-numbers

$$B_a = \begin{pmatrix} c\sigma p_a & mc^2 I_2 \\ mc^2 I_2 & -c\sigma p_a \end{pmatrix}, \tag{13.82a}$$

with the matrix I being of the order 4^n. Representing the coordinate wave function (multispinor) under the integral as a Fourier-transformation,

$$\Psi(\mathbf{r}) = \frac{1}{(2\pi)^{3n/2}} \int \exp\left(\sum_{k=1}^{n} i\mathbf{p}'_k\mathbf{r}_k\right)\Phi(\mathbf{p}')\, d^{3n}\mathbf{p}', \tag{13.82b}$$

changing the order of integration by the coordinates and momenta in the system Eq. (13.49), and defining the Fourier-transformation for the potential function

$$\bar{V}(\mathbf{p},\mathbf{p}') = \frac{1}{(2\pi)^{3n}} \int \exp\left(\sum_{k=1}^{n} -i\mathbf{p}_k\mathbf{r}_k\right)\left(\sum_{a=1}^{n} V_a + \sum_{a>b=1}^{n} V_{ab}\right)\exp\left(\sum_{k'=1}^{n} i\mathbf{p}'_{k'}\mathbf{r}_{k'}\right) d^{3n}\mathbf{r}, \tag{13.82c}$$

one obtains an integral equation system

$$\sum_{a=1}^{n} \mathrm{I}_4 \times \ldots \times \mathrm{B}_a \times \ldots \times \mathrm{I}_4 \cdot \Phi(\mathbf{p}) + \int \bar{V}(\mathbf{p},\mathbf{p}')\cdot \mathrm{I}\cdot \Phi(\mathbf{p}')\, d^{3n}\mathbf{p}' = E\Phi(\mathbf{p}). \tag{13.83}$$

The kinematic matrix in the left hand side of the equation can be represented as the spectral resolution into its eigenvectors as

$$\sum_{a=1}^{n} \mathrm{I}_4 \times \ldots \times \mathrm{B}_a \times \ldots \times \mathrm{I}_4 = \sum_{\nu=1}^{4^n} \mathbf{c}^*_\nu \lambda_\nu \tilde{\mathbf{c}}_\nu, \tag{13.84}$$

where $\lambda_v = \sum_{a=1}^{n} (-1)^{l(v,a)} \sqrt{m^2c^4 + c^2 p_a^2}$, and $l(a,v) = 1$ or 2, being dictated by the appropriate eigenvector. A real state of a particle is described by the arithmetic root, as the length of the eigenvector is 4^n. However, this is as demonstrated with the Dirac equation, in the transformation discussed of all the roots of the matrix (13.84) participating.

By virtue of great multiplicity of the eigen-numbers (it may be compared with composition of spins in atomic one-particle models) one obtains 2^n eigenvectors for the same eigenvalue as the arithmetic root. So this root gives the reasonable estimation for the electron system energy in an atom. All the rest, roots in which negative radicals enter, have no physical meaning, and they ought to be considered only like auxiliary algebraic constituents, while linearizing the kinetic energy operator of a particles system.

The Eq. (13.83) can be transformed to a new form, taking into account the relationship (13.84). Thus, in analogy with the Dirac equation, multiplying it to the left by the row eigenvector $\tilde{c}_1$ and using the orthonormality of the kinetic matrix eigenvectors, we arrive at the many-particle relativistic integral equation over functions χ_k.

$$\lambda_1 \chi_1(p) + \int \bar{V}(p,p') \times \tilde{c}_1(p) \sum_{k=1}^{4^n} c_k^*(p') \times \chi_k(p') d^{3n} p' = E \chi_1(p), \quad (13.85)$$

where $\lambda_1(\mathbf{p}) = \sum_{k=1}^{n} \sqrt{m^2c^4 + c^2 p_k^2}$. The rest $4^n - 1$ integral equations of the system have analogous forms. Each eigenvector can be written down in an explicit form, so an analysis of the kernel of the integral operator, which is cumbersome, can be made easily. As it may be seen, the kernel is factorized with respect to the function-components of the eigenvectors. Reduction of the kernel to a degenerate one, amounts to a factorization of the potential function by variables. This question demands a separate consideration. The property of the integral operator Eq. (13.85) may be noted. It is obvious, the Coulombic function of the kernel has the singularity when $\mathbf{p} = \mathbf{p}'$. The multiplier which includes the eigenvectors of the kinematic matrix, is represented by the sum in which the first adduct is reduced to unity, the other terms vanish because of eigenvector orthogonality, provided the momenta are equal. If the momentum is changed like the argument of an eigenvector, like the latter is rotated in many-dimensional vector space relative to the other eigenvectors with some other argument, as are the scalar products with them of the eigenvector $\mathbf{c}_1$ mean cosines of the angles among those vectors, the values of these are close to zero. In particular this note, concerns the eigenvectors which belong to roots with nonphysical negative energies.

Thus, the leading term in this kernel multiplier proves to be that with which the index coincides and with the index of the function outside of the integral operator. Besides, the Coulombic part of the integral operator kernel is singular, when the momenta are equal, and it defines the asymptotic behavior of the solution. In this case, the first term of the scalar products of the kinetic matrix eigenvectors only affects the wave function asymptotic, which is unity at equal arguments in the potential function. The rest scalar products are equal to zero because of the eigenvector orthogonality in this domain. Therefore one has a reason to solve first the scalar relativistic equation

$$\lambda_1 \chi_1(p) + \int \bar{V}(p,p') \times \tilde{c}_1(p) \sum_{k=1}^{4^n} c_k^*(p') \times \chi_k(p') d^{3n} p' = E\chi_1(p). \quad (13.86)$$

This equation can be solved by an iteration method choosing trial function from the decomposition of the potential function and components of the eigenvector $\mathbf{c}_1(\mathbf{p})$. A factorization of the Coulombic kernel $\bar{V}(\mathbf{p},\mathbf{p}')$ by variables gives an integral equation with a degenerated kernel for which solution can be obtained by an algebraic method. After calculation of the wave function in the momentum space, a transfer to the electronic coordinate space is made by the formula (13.81).

The eigenvectors corresponding to nonphysical roots of the Dirac equation (which are assigned conventionally to positron states) contribute just as algebraic elements of the relativistic model considered similarly the connection of real and complex roots of a polynomial with real coefficients. The prediction of the positron existence is truly connected not with the Dirac equation algebra, but with sign symmetry of the elementary electric charges, and this kinetic energy undoubtedly does not depend on an electric charge sign.

The Eq. (13.86) changes to the nonrelativistic Schrödinger equation for an atom if particle momenta are much lower than mc, then the eigenvector $\mathbf{c}_1$ has zero "positronic" components like in Eq. (13.37).

(viii)A hypercomplex representation in the momentum space of particles

The above-developed theory of relativistic equations for a heavy atom can be expanded to a hypercomplex variant of equations. We start from the equation system for an n-electron atom

$$\mathrm{H}\Psi = (E - V)\mathrm{I}\Psi, \tag{13.87}$$

where

$$\mathrm{H} = \mathrm{H}_1 \otimes \mathrm{I}_2 \otimes \ldots \otimes \mathrm{I}_2 + \mathrm{I}_2 \otimes \mathrm{H}_2 \otimes \ldots \otimes \mathrm{I}_2 + \ldots + \mathrm{I}_2 \otimes \mathrm{I}_2 \otimes \ldots \otimes \mathrm{H}_n, \tag{13.88}$$

with the kinematic matrices for separate particles

$$\mathrm{H}_k = \begin{pmatrix} \alpha\gamma p_{\Gamma k} & 1 \\ 1 & -\alpha\gamma p_{\Gamma k} \end{pmatrix}, \tag{13.89}$$

and the momentum written in the quaternionic form

$$p_\Gamma = ip_x + jp_y + kp_z, \tag{13.90}$$

with the Hamilton algebra $ijk = -1$, $i^2 = j^2 = k^2 = -1$, and $\alpha = \sqrt{-1}$ commuting together with the Hamilton units, $\gamma = e^2 / \hbar c$.

The atomic potential function is of the form (in the relativistic atomic scale)

$$V = \gamma^2 \sum_{q=1}^{n} -\frac{Z}{r_q} + \gamma^2 \sum_{q \neq q'=1}^{n} \frac{1}{r_{qq'}}, \tag{13.91}$$

Writing down the spectral resolution of the kinematic matrix (13.88)

$$\mathrm{H} = \sum_{k=1}^{2^n} \mathbf{c}_k^* \lambda_k \tilde{\mathbf{c}}_k \,. \tag{13.92}$$

in the momentum representation Eq. (13.87) then the integral equation is

$$\sum_{k=1}^{2^n} \mathbf{c}_k^* \lambda_k \tilde{\mathbf{c}}_k \Phi(\mathbf{p}) = E\Phi(\mathbf{p}) - \int d\mathbf{p}' \, V(\mathbf{p},\mathbf{p}')\Phi(\mathbf{p}'), \qquad (13.93)$$

where the potential function is expressed in the momentum space and defined with the help of the Fourier transformation of the corresponding potential function in the coordinate representation.

Multiplying this equation to the left successively by the eigenvectors of the kinematic matrix, one obtains equations with respect to projections of the multispinors onto those vectors. One gets

$$\lambda_k \psi_k(\mathbf{p}) = E\psi_k(\mathbf{p}) - \int d\mathbf{p}' \, V(\mathbf{p},\mathbf{p}')\mathbf{c}_k(\mathbf{p}) \cdot \mathrm{I} \cdot \Phi(\mathbf{p}'), \qquad (13.94)$$

where $\psi_k(\mathbf{p}) = \tilde{\mathbf{c}}_k(\mathbf{p})\Phi(\mathbf{p})$. For the following transformation of the equation, introducing the unit matrix of the order 2^n, and its spectral decomposition

$$\mathrm{I} = \sum_{k=1}^{2^n} \mathbf{c}_k^*(\mathbf{p}') \, \tilde{\mathbf{c}}_k(\mathbf{p}'), \qquad (13.95)$$

substituting it into the integral Eq. (13.94), one arrives at the equation system sought for the functions $\psi_k(\mathbf{p})$

$$\lambda_k \psi_k(\mathbf{p}) = E\psi_k(\mathbf{p}) - \int d\mathbf{p}' \, V(\mathbf{p},\mathbf{p}')\tilde{\mathbf{c}}_k(\mathbf{p}) \sum_{k'=1}^{2^n} \mathbf{c}_{k'}^*(\mathbf{p}')\psi_{k'}(\mathbf{p}'), \quad k = 1,2,\ldots,n, \qquad (13.96)$$

By an analogy with the system given in the preceding section, we can make the analogous conclusions concerning applications of this relativistic equation system in the theory of heavy elements.

The distinctive feature of the hypercomplex equation system, as compared with that considered in the Clifford space is that, among the eigenvectors of kinematic matrix there exists only one, corresponding to the sum of the positive roots of the quadratic equation for the free particle energy. So, the equation with the index one is a determining one for solving the physical problem on motion of n particles (electrons) near a force center. This equation can give the first approximation for the wave function of an atomic system. (It is possible a generalization of such an equation on molecular systems, also, but not in the context of this research.) This equation also gives the correct asymptotic solution of the initial equation system, with a singular kernel of the integral operator. Writing down the abovementioned scalar relativistic equation

$$\lambda_1 \psi_1(\mathbf{p}) = E\psi_1(\mathbf{p}) - \int d\mathbf{p}' \, V(\mathbf{p},\mathbf{p}')\tilde{\mathbf{c}}_1(\mathbf{p})\mathbf{c}_1^*(\mathbf{p}')\psi_1(\mathbf{p}'), \qquad (13.97)$$

where the scalar product of the multidimensional eigenvectors (spinors) of the kinematic matrix models is an influence of the spin kinematic of the particle system on their interactions by means of the Coulombic forces with the force center and between particles.

It stands to reason that advantages and deficiencies of these approaches given can be estimated in numerical realizations. We hope that the simple structure of the investigated equations will allow in the future to create mathematical software for posing and solving some actual problems in atomic spectroscopy and atomic physics of heavy and superheavy chemical elements.

KEYWORDS

- **Heavy atoms**
- **Hypercomplex algebra**
- **Many-particle relativistic quantum theory**
- **Momentum representation**
- **Quaternionic quantum models**

REFERENCES

1. Bethe, H. A.; and Salpeter, E. E.; Quantum Mechanics of One- and Two-Electron Atoms. Springer-Verlag: Berlin-Goettinger-Heidelberg; **1957**.
2. Fock, V. A.; Zur theorie des Wasserstoffatoms. *Z. Phys.* **1935,** *98,* 145.
3. Bateman, H.; and Erdélyi, A.; Higher Transcendental Functions. New York-Toronto-London: Mc Graw-Hill Book Company, Inc.; **1953,** *2.*
4. Novosadov, B. K.; Methods of solving the quantum chemistry equations. Foundations of the Molecular Orbital Theory. Moscow: Nauka; **1988,** 184 p. and 32 p. (in Russian)
5. Novosadov, B. K.; Methods of Mathematical Physics of Molecular Systems. Moscow: Book House "LIBROCOM"; **2010,** 383 p. and 73 p. (in Russian)

CHAPTER 14

A TECHNICAL NOTE ON IMPROVING ADHESION PROPERTIES OF RUBBERS

V. F. KABLOV, N. A. KEIBAL, S. N. BONDARENKO, D. A. PROVOTOROVA, and G. E. ZAIKOV

CONTENTS

14.1 INTRODUCTION

Today, despite the existence of a large number of adhesives, which differ not only in the composition and properties, but also in manufacturing technology, formulations, and intended purpose, the problem of creating new adhesives with a certain set of properties is still relevant. This is because of that fact that glue compositions are imposed higher demands related to operation conditions of construction materials and products.

The problem may be solved by applying the targeted modification to a film-forming polymer which is the base component of any glue composition. Modification is of a priority than creating completely new adhesive formulations. The modification process is more advantageous from both an economic and technological points of view and allows not only to improve the performance of rubber, but also to maintain a basic set of properties.

As it is known, there are several methods of film-forming polymer modifications. They are physical, chemical, photochemical modification, modification with biologically active systems, and a combination of all these methods.

Epoxydation, being one of the variants of chemical modification, represents a process of introduction of epoxy groups to a polymer structure thereby improving the properties of this polymer. Materials based on epoxidized polymers show high physical and mechanical characteristics, meet the requirements of the strength and dielectric parameters, and manifest as a good adhesion to metals, which is achieved as a result of high adhesive activity of epoxy groups. Owing to these properties they find application as coatings for metals and plastics, adhesives, mastics and potting compounds in electrotechnics, microelectronics, and other areas of engineering [1].

Chlorinated natural rubber (CNR), applied as an additive in glue compositions based on chloroprene rubber, is widely used in industry for gluing of different rubbers together with metals or with each other [2]. Individual brand glues based on CNR are rarely produced.

Isoprene rubber is an analog of natural rubber, used in the majority of rubber adhesives, but because of its low cohesive strength applied in their formulations much more rarely.

Therefore, the investigations concerned with the development of glue compositions based on these rubbers with improved adhesion characteristics to materials of different nature are of particular interest. It can be achieved by modification [3].

14.2 METHODOLOGY

In this work we investigated the possibility of chlorinated natural rubber's and isoprene rubber's epoxydation by means of ozonation. This was with the aim of improving adhesion characteristics of glues based on these rubbers as it is known that epoxy compounds are good film formers in glue compositions and increase the overall viscosity of the latter.

The introduction of epoxy groups in a polymer structure was carried out by means of ozonation as ozone is highly reactive toward double bonds, aromatic structures, and C-H groups of a macrochain.

The contact time (0.5–2 h) was varied in the ozonation process. The remaining parameters which are the ozone concentration (5×10^{-5}% volume) and temperature (23°C) were kept constant.

Further, glue compositions based on the ozonized rubbers were prepared.

Glue compositions based on the ozonized CNR were 20 percent solutions of the rubber in an organic solvent which was ethyl acetate. The composition based on the isoprene rubber was 5 percent solution of the ozonized rubber in petroleum solvent.

The gluing process was conducted at 18–25°C with a double-step deposition of glue and storage of the glued samples under a load of 2 kg for 24 h. Glue bonding of the vulcanizates was tested in 24 (±0.5) h after constructing a joint by method called "Shear strength determination" (State Standard 14759-69), in quality of samples using polyisoprene (SKI-3), ethylenpropylen (SKEPT-40), butadiene-nitrile (SKN-18), and chloroprene (Neoprene) vulcanized rubbers.

14.3 RESULTS AND DISCUSSION

In ozonation, a partial double bonds breaks in the rubber macromolecules leading to the macroradicals formation (Figure 14.1). Ozone molecules attaches at the point of the rubber double bonds breakage with the formation of epoxy groups [4]:

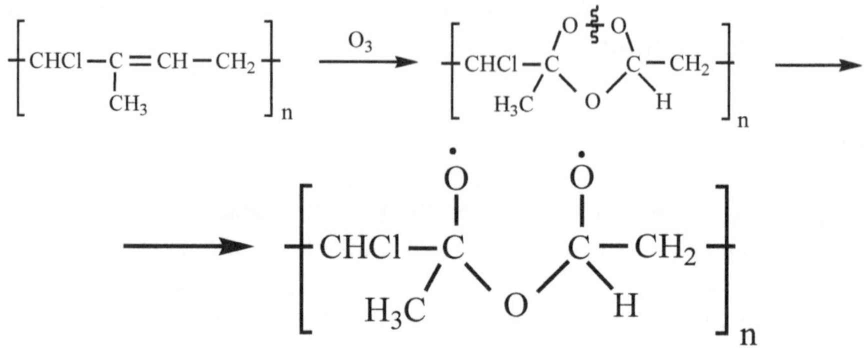

FIGURE 14.1 Reaction of the macroradicals formation (on the example of CNR).

The formed macroradicals are probably interacting with macromolecules of the rubber which is the substrate material, thereby increasing its adhesion strength.

Initially, the CNR of three brands was investigated, namely S-20, CR-10, and CR-20. The results obtained in ozonation of three rubber brands are shown in the Figure 14.2.

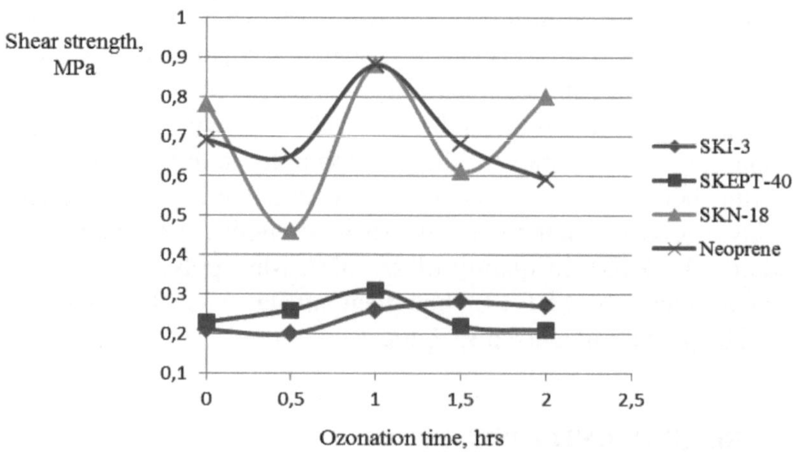

(a)

(a)

FIGURE 14.2 *(Continued)*

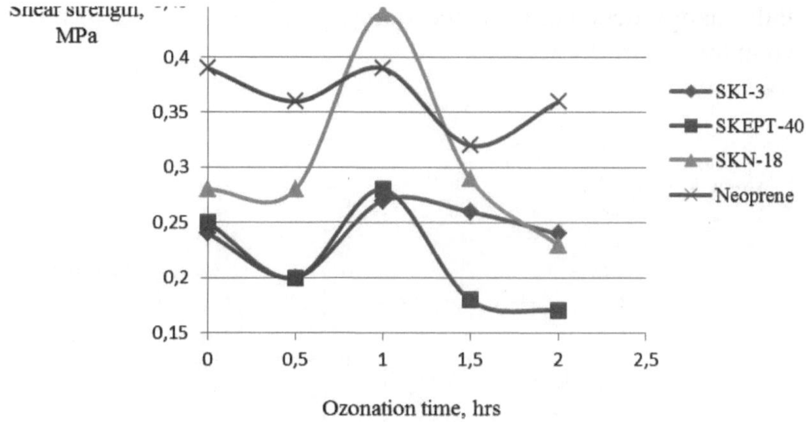

(b)

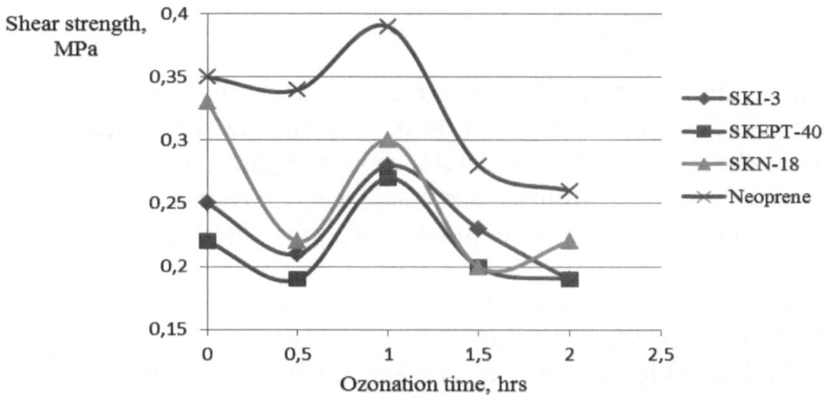

(c)

FIGURE 14.2 Influence of ozonation time on shear strength at gluing of vulcanizates with glue compositions based on CNR of the S-20 (a), CR-10 (b) and CR-20 (c) brands respectively.

A decrease in the adhesion strength at $\tau = 0.5$ h may be connected with preliminary destruction of macromolecules under action of reactive ozone.

From the Figure 14.2 we can see that the maximal figures correspond to 1 h ozonation. The adhesion strength for rubbers based on different caoutchoucs increases by 10–40 percent at that. The results of the shear strength change depending on the rubber brand and adherend type are shown in the Figure 14.3.

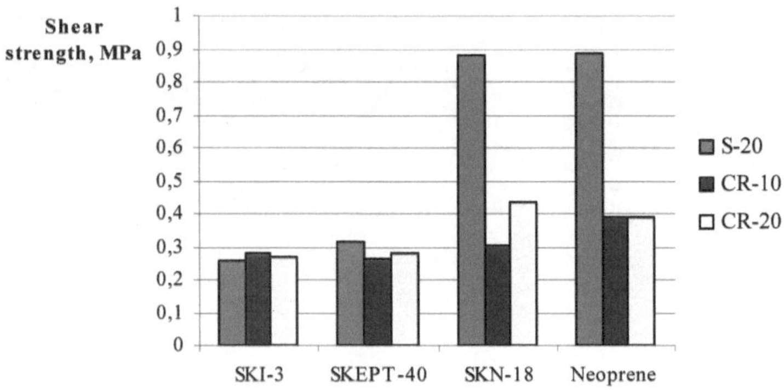

FIGURE 14.3 A change in shear strength for different CNR brands depending on adherend type (ozonation time $\tau = 1$ h).

It should be noticed that the extreme nature of the above-mentioned dependences can be explained by the diffusion nature of the interaction between adhesive and substrate. As shown in the figures, with the increasing content of functional groups the strength began to reduce, having reached a certain limit in the adhesive. In this case, only adhesive molecules have ability to diffusion [5].

Isoprene rubber was also treated with ozone at the same parameters maintained for CNR. The results are on the Figure 14.4.

The data in Figure 14.4 confirm the ambiguity of the ozonation process. When the contact time is equal to 15 min (as in the case of CNR epoxydation [6]), possible preliminary destruction of the rubber macromolecules takes place, which on the graphs is proved by almost simultaneous reduction in the shear strength values. Concurrently, formation and subsequent growth of macroradicals go on, as evidenced by the increase in adhesive characteristics at ozonation time 0.5 and 1 h. Here the shear strength at gluing of different vulcanized rubbers increases by 10–70 percent on average, and then it starts to reduce again.

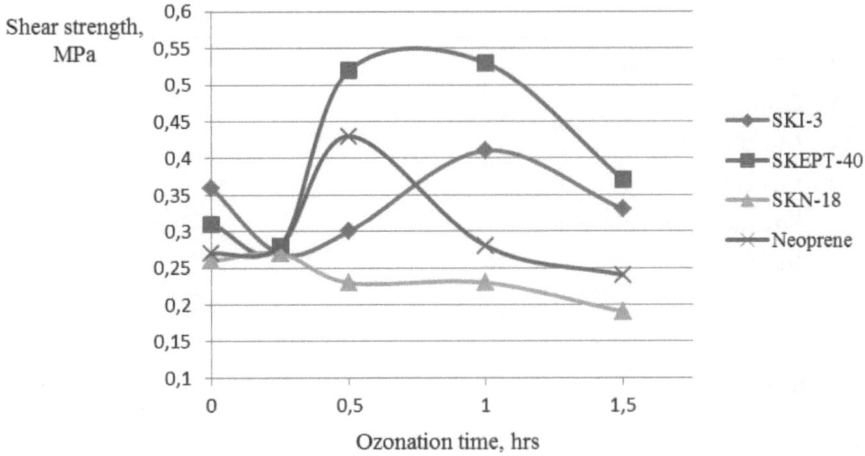

FIGURE 14.4 Influence of ozonation time on adhesive strength for the compositions based on isoprene rubber.

With further increase in ozonation time, the values of adhesion strength decline. That is apparently related to saturation of the polymer chain with epoxy groups and decrease in mobility of the macromolecules, and consequently, the degree of interaction of the substrate with the adhesive composition as well as with destruction of the polymer chains.

14.4 CONCLUSIONS

Thus, ozonation can be applied as an effective method for enhancing adhesion properties of rubbers with modification of film-forming polymers that are the main component in glues. Changing one of the parameters during the ozonation process we can obtain such a content of epoxy groups at which the characteristics of adhesion strength will be maximal.

KEYWORDS

- **Glue compositions**
- **Gluing**
- **Modification**
- **Ozonation**
- **Unsaturated rubbers**
- **Vulcanizates**

REFERENCES

1. Solovyev, M. M.; Local dynamics of oligobutadienes of different microstructure and their modification products: Thesis of Ph.D. in Chemistry Sciences: 02.00.06. Solovyev Mikhail Mikhailovich. Yaroslavl; **2009,** 201 pp.
2. Dontsov, A. A.; Ya, G.; and Lozovick, S. P.; Novitskaya. Chlorinated Polymers. Moscow: Khymiya; **1979,** 232 pp.
3. ablov, V. F.; Bondarenko, S. N.; and Keibal, N. A.; Modification of Elastic Glue Compositions and Coatings with Element Containing Adhesion Promoters: Monograph. Volgograd: IUNL VSTU; **2010,** 238 p.
4. Zaikov, G. E.; Why do polymers age. *Soros Educ. J.* **2000,** 6(12), 52.
5. Berlin, A. A.; and Basin, V. E.; Basics of Polymer Adhesion. Moscow: Khymiya; **1969,** 320 p.
6. Keibal, N. A.; Bondarenko, S. N.; Kablov, V. F.; and Provotorova, D. A.; Ozonation of chlorinated natural rubber and studying its adhesion characteristics. Rubber: Types, Properties and Uses. Popa, Gabriel A.; ed. New York: Nova Publishers; **2012,** 275–280 pp.

CHAPTER 15

A SHORT NOTE ON FIRE RESISTANCE OF EPOXY COMPOSITES

V. F. KABLOV, A. A. ZHIVAEV, N. A. KEIBAL, T. V. KREKALEVA, and A. G. STEPANOVA

CONTENTS

15.1 INTRODUCTION

Today, expanding the range of technological and operational characteristics of composite materials based on epoxy oligomers is an urgent problem. Polymer composites based on epoxy resins are widely used as structural materials and adhesives. The advantages of epoxy composites are good adhesion to reinforcing elements, the lack of volatile by-products during hardening, and low shrinkage [1, 2].

However, in some cases, the use of epoxy composites is limited by their low thermal stability and fire resistance [3]. One of the benefits of epoxy resins is the ability to regulate their composition by introducing various modifiers (fillers, plasticizers, fire retardants, etc.) resulting in materials with a given set of properties [4]. There are known fire retardant polymer compositions with microencapsulated fire extinguishing liquids (halogen and phosphorus containing compounds, water, etc.). Microencapsulation substantially improves both technological and functional properties of the most diverse products, and this considerably expands the scope of their application.

15.2 EXPERIMENTAL PART

In this research, the influence of hydrophilic filler content on the fire resistance of composites based on epoxy resin ED-20 was investigated.

Acrylamide copolymer POLYSWELL was applied as a hydrophilic filler. The acrylamide copolymer is a white granular material, with density of 0.8–1.0 g/cm^3, and swells in water to form a polymer gel. In solutions the amide group shows weak-basic properties at the expense of the lone electron pair on the nitrogen atom, and hence the reason for nonchemical interaction of the polymer with water.

The compositions were obtained on the basis of epoxy resin, by means of blending this with other components as follows: epoxy resin ED-20, –filler-acrylamide copolymer in the form of granules preliminary swollen in water to the ratio of 1:10, and hardener polyethylenepolyamine. The obtained reactive blends were molded and then hardened without heat supply for 24 h. The test samples had the following sizes: diameter of 50 mm, thickness of 5 mm.

For determining the efficiency of the developed composites, experiments on the fire resistance were conducted by exposure of a sample to open flame using the universal Bunsen burner. With a help of the pyrometer C-300.3, measuring the moment of achieving the limit state, the time-temperature transformations on the nonheated surface of the sample were registered.

Along with the fire resistance estimate of the developed compositions, studies were also carried out on samples for water absorption and combustibility depending on the hydrophilic filler content.

The combustibility was evaluated by the standard technique on the rate of horizontal flame spread over the surface. A sample was exposed to the burner flame (temperature peak 840°C) with fixed burning and smoldering time after fire source elimination.

The experiments on the water absorption were performed in distilled water at temperature 23 ± 2°C for 24 h. The water absorption was characterized by the sample weight change before and after exposure to water.

15.3 RESULTS AND DISCUSSION

As was mentioned above, the investigation on determining the fire resistant properties of epoxy compositions was carried out during the research. The results are as shown in Figures 15.1 and 15.2.

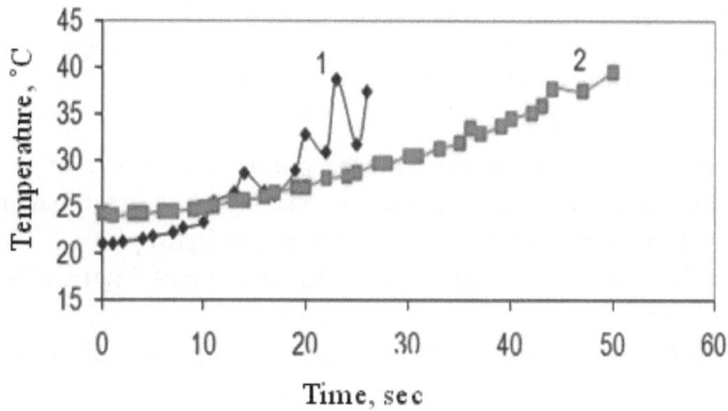

FIGURE 15.1 Dependence of temperature on the nonheated sample side on flame exposure time: 1—initial epoxy component; 2—epoxy composite containing hydrophilic filler.

As it was seen in Figure 15.1, the test sample damage took place at the 15th sec which was evidenced by temperature fluctuations on curve 1. The filled sample (containing 15% of the hydrophilic filler) maintains the integrity up to 50 sec; the sparking observed at combustion was probably related to water injection into the combustion zone. Besides, when the flame source was eliminated, the sample self-extinguished for 2–3 sec.

The effect of hydrophilic filler content (5–20%) on the fire resistance of the composites was also studied in this work (Figure 15.2).

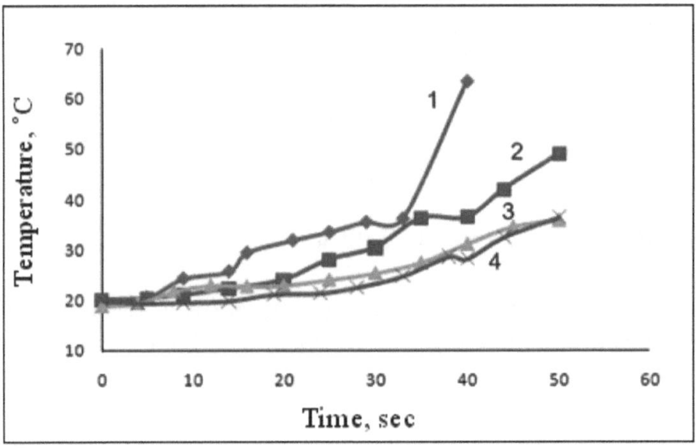

FIGURE 15.2 Dependence of temperature on the nonheated sample side on flame exposure time for epoxy composite containing hydrophilic filler in the following amounts: 5 percent (1), 10 percent (2), 15 percent (3) 20 percent (4).

When measuring temperature on the nonheated surface of water-containing composites within a specified time span, it was established that the fire resistant properties improve with increasing hydrophilic filler content from 10 to 20 percent; the sample with filler content of 5 percent was damaged by 33 sec.

The results of experiments on the water absorption and fire resistance are presented in Table 15.1.

TABLE 15.1 Estimation of water absorption and fire resistance of epoxy composites

Parameter	Filler Content in a Composite, % by Epoxy Resin Weight				
	Without Filler	5	10	15	20
	Composition Index				
Water absorption (%)	0.41	0.33	0.24	0.28	0.28
Time to the limit state (sec)	15.0	35.0	50.0	50.0	50.0
Temperature of the nonheated sample side in 25 sec (°C)	Sample damaged	33.0	28.0	24.0	20.0

Proposed compositions provided a significant increase in values of the fire resistance and water absorption compared to the initial sample. Time to the limit state for the test samples goes up by 2.5 times to that of the initial sample.

The data obtained in combustibility tests of water containing compositions are illustrated in Figure 15.3. The best results were obtained with filler content of 15 and 20 percent. In this way, the flame spread rate for the initial sample was 18 mm/min and, with the hydrophilic filler was equal to 3 mm/min.

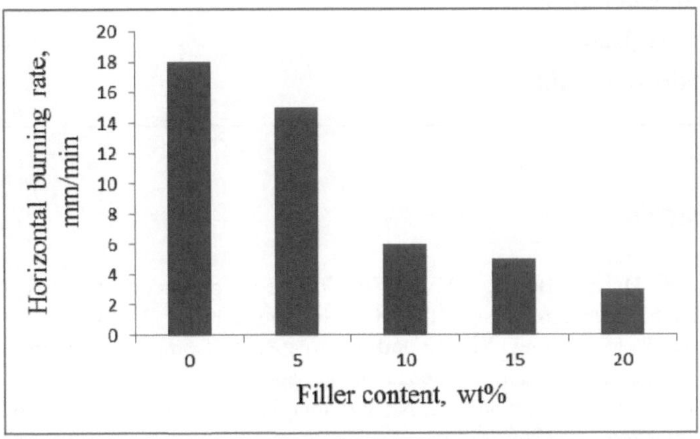

FIGURE 15.3 Evaluation of the horizontal flame spread rate over epoxy composite surface.

On exposure to flame, there was a kind of microexplosions with injection of fire extinguishing liquid–water–occurring in the combustion zone. In this case, combustion inhibition was likely due to the absorption of a significant heat amount, typical of the high heat capacity and high water evaporation heat. A possible factor in reducing the flame spread rate would be by water displacement of the combustion reaction components from the reaction zone.

15.4 CONCLUSION

So, the properties of water containing epoxy composites were developed and investigated as well as the possibility of applying a hydrophilic filler as an additive that increase the fire resistance of hardened epoxy compositions based on epoxy resin ED-20 was shown in this paper.

One of the important conclusions in this work is the high prospect of the modification methods for epoxy composites in order to give them a set of specific properties using filled microencapsulated materials.

KEYWORDS

- **Acrylamide copolymer**
- **Epoxy composites**
- **Fire resistance**
- **Hydrophilic filler**

REFERENCES

1. Eselev, A. D.; and Bobylev, V. A.; State-of-the-art production of epoxy resins and adhesive hardeners in Russia. Adhesives. Sealants, 2006, 7, 2–8 pp.
2. Amirova, L. M.; Ganiev, M. M.; and Amirov, R. R.; Composite Materials Based on Epoxy Oligomers. Kazan: Novoe znanie plc; 2002, 167 p.
3. Kopylov, V. V.; and Novikov, S. N.; Polymer Materials of Reduced Combustibility. Moscow: Khimiya; 1986, 224 p.
4. Kerber, M. L.; and Vinogradov, V. M.; Polymer Composite Materials: Structure, Properties, and Technology. Berlin, A. A.; ed. Professiya: St. Petersburg; 2008, 560 p.

CHAPTER 16

A COMMENTARY ON STRUCTURAL TRANSFORMATIONS OF PIRACETAM UNDER LEAD ACETATE INFLUENCE

O. V. KARPUKHIN, K. Z. GUMARGALIEVA, S. B. BOKIEVA, and A. N. INOZEMTSEV

CONTENTS

Piracetam (PIR), a standard representative of class of psychotropic medicine–nootrops [1], has been extensively used recently both in clinical practice and experimental studies. Piracetam features diverse functional properties and has ambiguous effect on learning capability and memory, even when introduced in small doses in various test series [2, 3].

It is natural to attribute this ambiguity to some unaccounted physicochemical factors. Among them, we can mention different contents that arise due to the medicine's conformers–like the polymorphic crystalline structures, which can give rise to various permolecular structures in PIR solutions [4].

The PIR's diverse effects can also be brought in by physicochemical characteristics of water. In this respect, of great interest are salts of heavy metals, for example, lead acetate, contained in water, including drinking water not purified with biological filters, and in appreciable quantities. If this supposition is true, in regions with a high metal content in water, one can expect unusual effects of PIR on superior neural human activity because of its extensive application in clinical practice.

The aforesaid stimulated us, on the one hand, to study the effect of PIR dissolved in water containing lead acetate on leaning and memory of rats, and also on the other to ascertain the changes in the PIR structure brought about by lead acetate additives.

We used 45 white strain-free male rats 180–200 g in weight divided into four groups: we introduced intraperitoneally PIR in amount of 300 mg/kg in rats of group I half an hour before each test, lead acetate solution (10^{-7} mol/L) in rats of group II 4.5 h before test, lead acetate solution and PIR in rats of group III, and physiological solution in rats of group IV (reference group). Then, for 5 days we developed (25 trials per day) an "avoidance" reflex in a shuttle chamber. Tests were carried out as follows: we switched on an irritant (800 Hz sound) and then 10 sec of electric current. With running of animals to another half of the chamber, switched off both stimuli. The pause between signals was 30–60 sec.

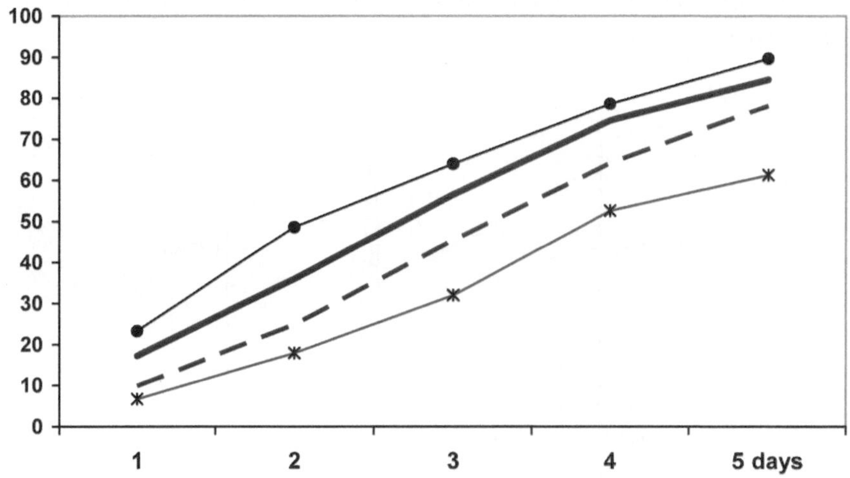

FIGURE 16.1 Dynamics (%) of avoidance of electric shock by rats injected with aqueous solutions if piracetam, lead acetate, and their mixture: —•— PIR ($p < 0.001$), □- - - - - Pb acetate, — reference, —ж— PIR + Pb.

To study structural changes, we prepared 2 percent (weight) PIR solutions in distilled water (solution I) and a 10^{-7} mol/L lead acetate solution in distilled water (solution II). Changes in the state of the solutions were analyzed by liquid chromatography.

Chromatograms are taken on a Beckman chromatograph with reversed ODS phase, isocratic elution (1 L/min), methanol and water (15:85 volume ratio) as eluents, and detection at a wave length equal to 220 nm. Calorimetric measurements were conducted on a Mettler—3301 differential scanning calorimeter (DSC) at a heating rate of 1°/min.

The experimental data displayed in Figure 16.1 suggest that neither PIR nor lead acetate introduced separately suppresses learning. Moreover, the learning curve for animals injected with PIR always goes above the reference, although this surplus attains a statistically significant level only on the 3rd day ($p = 0.01$). However the combined effect of these solutions suppresses learning, which shows up as a statistically significant reduction of avoidance reactions in the second test. These agents exerted a similar effect on the locomotion activity in the form of intersignal reactions, that is, individually they did not suppress this activity, but together they did.

A representative chromatographic trace of a PIR solution is show in Figure 16.2.

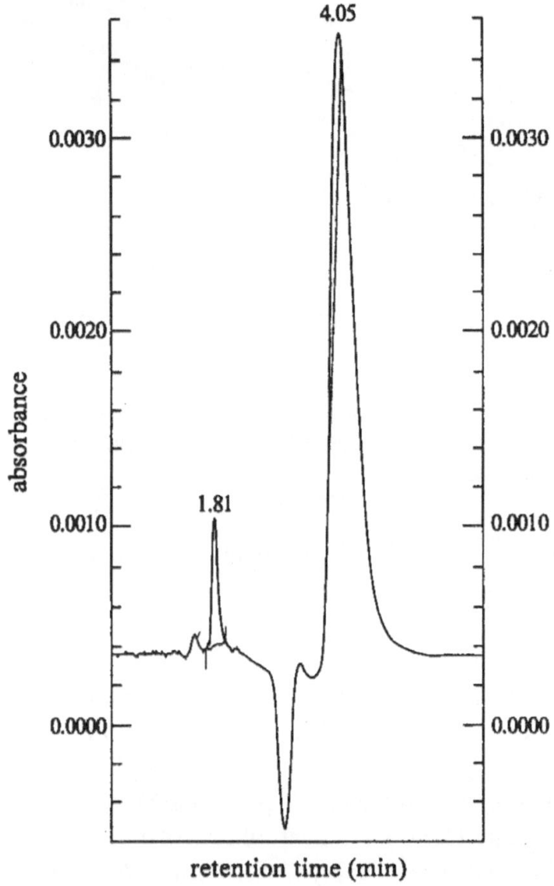

FIGURE 16.2 Chromatographic record of a newly prepared aqueous solution of Piracetam.

Lead acetate additives and solution storage do not change the signal retention time τ, it solely affects the areas under the appropriate curves. To get the chromatographic peaks, we ascertained the dependence of phase transition heats on chromatographic records of aqueous solution of PIR and the type of its thermal pretreating (Table 16.1).

The measurements showed that the medicine used features of its two melting points (400 and 420 K). This is supported by the data on the polymorphic structure of crystalline PIR [4].

Changes in the phase transition heat entailed variations of the UV signal in chromatographic analysis of aqueous solution of PIR. Changes in the heat at 400 and 420 K are correlated with amplitude of signals τ = 1.81 min and τ = 4.05 min respectively. It means, that piracetam is an association of different crystalline structures. Inasmuch as the temperatures at which the DSC curve exhibits an endothermic peak, they are independent of thermal pretreatment modes and the PIR sample weight does not change or decrease in the total phase transition heat (Table 16.1). At a constant specific heat, it implies that the amount of crystalline PIR in samples reduces as a results of thermal treatment.

This also signifies that along with the crystalline phase, solid PIR contains also a noncrystalline (amorphous) phase. When one of the crystalline structures is completely transformed into the amorphous phase (Table 16.1, line 2), the appropriate endothermic peak in the DSC record disappears, however the appropriate signal in the UV range partially remains. This suggests that solution of the amorphous PIR phase also absorbs in the UV range, though with a lower (about 30 time as low) extinction coefficient. The contribution of amorphous PIR to the absorption in the UV range is the major reason why a relative reduction of the UV signal upon one thermal treatment to another slightly differs (by 2–10%) from that of the reduction of melting heats under the same conditions.

Acetamide dimer's presence, as Figure 16.3 shows, solely affecting the crystalline state, with a melting point of about 400 K, can serve as these structure elements.

The results of chromatographic measurements with changes occurring in aqueous solutions at various times lapsed since the preparation of solution are summarized in Table 16.2. Both in solutions with lead acetate additives and with PIR proper the peak areas at τ = 1.81 min diminish, although at different rates, to values close to the appropriate values recorded in newly prepared solutions of amorphous PIR. The peak at τ = 4.05 min in solution without lead acetate is stable, while the peak area of solutions without lead acetate slightly rises with time.

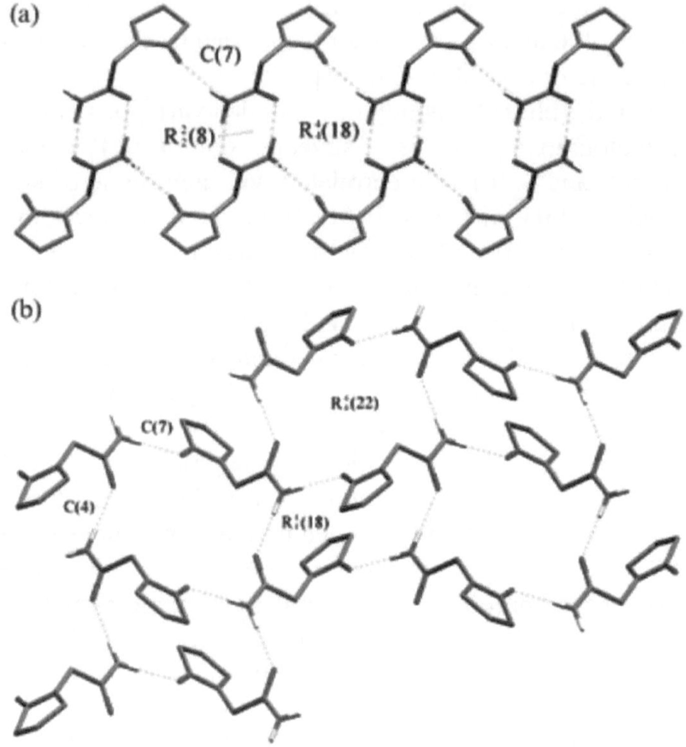

FIGURE 16.3 The molecular packing arrangements in the crystal structures of piracetam [4]. (a) For the melting point of about 420 K. Acetamide dimers are framed and (b) Is the structure pertaining to the polymorphic phase with a melting point of 400 K.

A decrease in the signal at $\tau = 1.81$ min signifies disturbance of the initial structure. The different behavior of the peaks at $\tau = 4.05$ min is associated with the fact PIR even in its amorphous state retains some structure elements in the absence lead acetate. These residual structure elements preclude formation of stable associates with structures inherent in $\tau = 4.05$ min.

Lead acetate present in solution II is capable of producing strong complexes with amide groups and destroys acetamide dimers facilitating thereby transition from one structure to another. The above-mentioned "suppression" effect is presumably associated with a significant increase in the content of the biologically active dissociation product of the aforesaid dimer in the solution.

TABLE 16.1 The effect of thermal pretreating mode on phase transition heats of solid Piracetam and chromatographic records of newly prepared aqueous solutions

N/N	The Sort of Heat Treatment	Phase Transition Heat (J/g, at Temperature, K)		Chromatogram Peaks Area (Relative Units) at the Output Time (min)	
		400	420	1.81	4.05
1.	None	24.4	182.2	6.52	382.5
2.	Heating up to 475 K and cooling up to 300 K	0	170.5	0.189	351.0
3.	Heating up to 475, cooling up to 300 K and 2 weeks standing	9.2	172.0	2.98	355.0
4.	Heating up to 525 K and cooling up to 300 K	13.4	110.0	4.06	240.8

TABLE 16.2 The dependence of changing of chromatogram peaks upon the time from the moment of piracetam solutions preparations II and I.

Time from the Moment of Solution Preparation (min)	Area of Peak (Relative) Units for Two Output Time (min)			
	Solution I		Solution II	
	1.81	4.05	1.81	4.05
0	6.52	382.5	6.45	377.7
15	4.12	382.5	0.72	378.1
33	1.05	382.5	0.29	382.27
66	0.41	382.5	0.21	389.99
101	0.22	382.5	0.12	389.99
143	0.22	382.5	0.12	395.13

KEYWORDS

- **Lead acetate**
- **Piracetam**
- **Polymorphism**
- **Structural transformations**

REFERENCES

1. Giurgea C. E.; The nootropic concept and its prospective implications. *Drug. Dev. Res.* **1982,** *2,* 441–446.
2. Burov, Yu.; Inozemtsev, A.; Pragina, L.; Litvinova, S.; Karpukhina, O.; and Tushmalova, N.; *Bull. Exper. Biol. Med.* **1993,** *115(2),* 150–152.
3. Inozemtsev, A.; Kapitsa, I.; Garibova, T.; Bokieva, S.; and Voronina, T.; *Mos. Un. Biological Sci. Bull.* **2004,** *3,* 24–26.
4. Fabbiani, F. P. A.; Allan, D. R.; Parsons, S.; and Pulham, C. R.; *Cryst. Eng. Comm.* 2005, *7,* 179–186.

CHAPTER 17

SIMULATION IN THE LABORATORY CONDITIONS OF AEROBIC–ANAEROBIC BIOREMEDIATION OF OIL-POLLUTED PEAT FROM RAISED BOGS

SERGEY N. GAYDAMAKA and VALENTINA P. MURYGINA

CONTENTS

17.1 INTRODUCTION

The main oil production areas in Russia are situated in the Northern Siberia, and in the same places there are the most extensive bogs polluted with oil. Application of remediation technologies, developed in Russia, on impassable bogs polluted with oil is almost impossible technically, and economically unfavorable. Besides the severe climate with cold and long winters and short cool summers worsened by absence of any roads in tundra and forest-tundra, and emergency oil spills on fenny bogs, makes it impassable for special machinery devices.

Therefore an elimination of such spills and their consequences on the bogs is the actual and difficult problem there. Depth of oil penetration on these bogs does not exceed of 0.6–1.0 m and often is propped up with water or permafrost. Processes of self-restoration of such bogs can take several hundred years. The pollution can extend in width there, and hence causing irreparable damage to the nature of the Polar Region.

In 2011, there was an attempt to clean oil in a strongly polluted bog using an augmentation of bacterial oil-oxidizing preparation Rhoder. Oil spill was in spring, and the oil was partially collected with a pump from the sludge. The bog was then watered three times with the Rhoder and one time with a fertilizer and lime. As a result level of oil pollution in the peat was decreased by 32–98 percent depending on initial concentration of oil which varied from 21–29 kg of crude oil on 1 kg of absolutely dry matter (DM) to 450–850 g/kg DM, and a depth of penetration of oil into the moss. The received results have induced a development of a new remediation technology in laboratory conditions with use of electron acceptors and the Rhoder to enhance the oil oxidation on the surface and in the depth of the peat. In this paper, there is a presentation of an attempt to develop a new aerobic–anaerobic bioremediation technology for fenny bogs polluted with oil for using it in the northern part of Russia.

17.2 MATERIALS AND METHODS

The microbial oil-oxidizing preparation Rhoder was used in laboratory experiment. The Rhoder consists of two bacterial strains Rhodococcus (R. ruber Ac-1513 D and R. erythropolis Ac-1514 D) picked out from soils, polluted with oil. Strains were neither pathogenic to people, animals, and plants, nor caused mutations in bacteria. This Rhoder is allowed for broad use in the nature. It was successfully applied to bioremediation of oil sludge, soils, bogs, and surfaces of water from oil pollution [1–8].

Installation was made through vertical plastic pipes (five models) with a length of 100 cm and diameter of 10 cm attached to a board (Figure 17.1). In each model two openings with a diameter of 2 cm at distance of 40 and of 90 cm from the upper edge were made for sampling. Each model was filled with the natural peat polluted with oil with a high concentration of hydrocarbons (HC) from 370 g/kg to 550 g/kg of DM.

17.2.1 SCHEME OF THE EXPERIMENT

- Model No. 1—negative control in which was added water for maintenance of high humidity of the peat, which was typical for bogs.
- Model No. 2—activation of indigenous microorganisms with mineral fertilizers that were added into the top layer of the peat up to the depth of 10 cm, and introduction of a gaseous electron acceptor up to the depth of 40 cm from the top layer of the peat in the model.
- Model No. 3—processing of the top layer of the peat up to the depth of 10 cm with the water solution of the Rhoder and fertilizers, and injection of the gaseous electron acceptor up to the depth of 40 cm from the top layer of the peat in the model.
- Model No. 4—processing of the top layer of peat up to the depth of 10 cm with the water solution of the Rhoder and fertilizers, and liquid electron acceptor on the top layer of the peat in the model.
- Model No. 5—processing of the top layer of the peat up to the depth of 10 cm with the water solution of the Rhoder and fertilizers and injection of the liquid electron acceptor up to the depth of 40 cm from the top layer of the peat in the model.

17.2.2 CARRYING OUT BIOREMEDIATION

Soils in the models No 3–5 were processed three times with working solution of the Rhoder with a number of hydrocarbon oxidizing cells (HCO) of 1.0×10^8 cells/mL by watering with an interval of 3 weeks. The fertilizer («Azofoska» C:N:P 16:16:16) was used, and 40 mL of the solution was added to the models three times. The gaseous and liquid electron acceptors were used and injected into models, according to the scheme of the experiment. The top layers of the peat in all models were maintained on a humidity of no less than 60 percent, and the top layers of the peat were mixed two times a week and before each introduction of fertilizer and the Rhoder.

17.2.3 SAMPLING

Soils sampling from models were made before the experiment beginning and before every application of the fertilizer and the Rhoder and each injection of electron acceptors which were entered into models. Samples from models were selected from the depth of 0 to 10 cm, 40 cm, and 90 cm from the upper edge of each model, for conducting chemical, agrochemical and microbiological analyses.

17.2.4 CHEMICAL AND AGROCHEMICAL ANALYSES

Oil in each sample of the dry peat was extracted on a Sockslet device with boiling $CHCl_3$, and its quantity gravimetrically determined. Then each dry material extracted by chloroform was fractioned on a mini-column with silica gel (Diapak-C). Oil products were analysed by the gas chromatograph (GC). GC model was the KristalLuks 4000м (by company Metakhrom) with the NetChrom V2.1 program, the column OV-101 length of 50 m, internal diameter of 0.22 mm, thickness of the phase of 0.50 μ, the FID detector, the temperature of the detector 300°C, the evaporator temperature of 280°C, the gradient from 80°C to 270°C, the velocity of raising temperature was 12°C/min. [9].

pH of each sample, humidity, and the general maintenance of the available nitrogen and phosphorus were determined with colorimetric methods [10].

17.2.5 MICROBIOLOGICAL ANALYSES

Most probable number (MPN) of microorganisms was determined by using tenfold dilutions and cultivation on meat-peptone agar in Petri dishes and using of selective agar nutrients for identification of ammonifying microorganisms: Actinomycete, Pseudomonas, oligotrophic bacteria, and

Micromycete. Most probable number of anaerobic microorganisms (mainly sulfate-reducing bacteria (SRB)) in samples of the peat which have been selected from the depth of 40 and 90 cm from models, were determined on the liquid Postgate medium [11].

Determination of MPN of oil-oxidizing microorganisms in samples of peat was made with use of modified liquid Raymond's media with oil as a sole carbon source (g/L): Na_2CO_3-0.1; $CaCl_2$.6 H_2O-0.01; $MnSO_4$.7 H_2O-0.02; $FeSO_4$-0.01; Na_2HPO_4.12H_2O-1.0; KH_2PO_4-1.0; $MgSO_4$.7 H_2O-0.2; NH_4Cl-2.0; NaCl-5.0; pH = 7.0 [12].

17.3 RESULTS AND DISCUSSION

Laboratory experiment was performed on the models which imitated an over-wetted bog polluted with oil. Preliminary microbiological analyses of samples taken from the top, middle, and bottom layers of the peat on the length of models showed that the peat in all models had different species of microorganisms: Bacillus, Pseudomonas, Rhodococcus, SRB, and Penicillium. In the top layers of the peat (0–10 cm) in each model they discovered the MPN of heterotrophic bacteria (HT) from 6.0×10^7 to 1.1×10^8 c.f.u./g of the peat, HCO bacteria from 9.1×10^5 to 9.4×10^6 cells/g of the peat. The anaerobic microorganisms were not determined in the top layers of the peat in each model. In samples of the peat, which have been selected from the middle parts of the models, the MPN of HT varied from 8.1×10^5 to 3.7×10^7 c.f.u./g of the peat, MPN of HCO bacteria varied from 8.2×10^2 to 9.8×10^4 cells/g of the peat. MPN of anaerobic microorganisms (SRB) 1.0×10^2 cells/g of peat were found in the samples from the middle of models No 3 and 4. In the bottom samples of the peat in these models, there were found anaerobic and microaerophilic bacteria with MPN from 2.1×10^6 to 4.9×10^7 c.f.u./g of the peat, and HCO bacteria from 7.1×10^3 to 1.0×10^6 cells/g of the peat. SRB were found in the bottom samples from the models No 2, 3, and 5 with the MPN from 1.0×10^2 to 1.0×10^5 cells/g of the peat (Table 17.1).

Agrochemical analyses show that the content of nitrogen and phosphorus in the peat in models were rather high and the ratio of C:N:P was in average 100:0.1:0.01 (Table 17.1).

TABLE 17.1 Microbiological and agrochemical characteristics of peat samples from the different length of the models before bioremediation

No Model	Point of Sampling	pH	HT cfu/g of Peat	HCO, Cells/g of Peat	SRB, Cells/mL	$N\text{-}NH_4^+$ mg/kg of Peat	PO_4^{3-} mg/kg of Peat
1	Top	5.9	$8.5*10^7$	$9.4*10^6$	-	507.13	459.7
	Middle	6.2	$8.1*10^5$	$8.1*10^2$	**0**	516.2	393.6
	Bottom	6.7	$2.9*10^6$	$9.7*10^4$	**0**	544.1	418.3
2	Top	5.8	$8.3*10^7$	$9.4*10^6$	-	450.8	393.6
	Middle	6.4	$3.6*10^7$	$9.8*10^4$	**0**	453.7	471.1
	Bottom	6.1	$1.4*10^7$	$7.1*10^3$	$1,0*10^5$	495.2	318.6
3	Top	6.0	$1.1*10^8$	$1.1*10^5$	-	427.3	402.6
	Middle	6.2	$3.7*10^7$	$8.7*10^4$	$1,0*10^2$	440.7	401.1
	Bottom	6.3	$4.9*10^7$	$1.1*10^4$	$1,0*10^2$	428.8	370.2
4	Top	6.3	$1.1*10^8$	$9.1*10^4$	-	402.5	391.2
	Middle	6.3	$6.9*10^6$	$1.1*10^4$	$1,0*10^2$	448.6	411.9
	Bottom	6.1	$2.8*10^6$	$9.4*10^4$	**0**	463.9	506.7
5	Top	6.0	$6.0*10^7$	$9.5*10^5$	-	494.7	329.7
	Middle	6.2	$3.4*10^7$	$1.1*10^4$	0	454.1	278.1
	Bottom	5.9	$2.1*10^6$	$1.0*10^6$	$1,0*10^5$	454.7	364.2

Note: Not detected.

At completion of the experiment, it was noted the MPN of microorganisms (HT and HCO bacteria) grew on 1 or 2–3 orders practically in all models in the top layers of the peat, and decreased on 1–2 orders in the middle and bottom layers of the peat. At the same time the number of anaerobic microorganisms including SRB in all models significantly grew in the middle and the bottom layers of the peat that can be connected with formation of own biocenosis in each model (Table 17.2).

Concentration of biogenic elements in all layers of models changed and even decreased and this can be connected with activation of microorganisms in each model. In the bottom and in the middle parts of the models concentration of biogenic elements decreased and this can be connected with activation of anaerobic microorganisms in each model (Table 17.2).

TABLE 17.2 Microbiological and agrochemical characteristics of peat samples from the different length of the models after the end of the bioremediation

No Model	Point of Sampling	pH	HT CFU/g of Peat	HCO, Cells/g of Peat	Other anaerobic Bacteria/ SRB, Cells/ml	N-NH_4^+ mg/kg of Peat	PO_4^{3-} mg/kg of Peat
1	Top	5.8	5.7*10^7	1.1*10^7	-	316.9	168.9
	Middle	5.7	2.6*10^8	9.9*10^1	0/5*10^3	425.8	276.7
	Bottom	5.2	3.4*10^7	8.2*10^3	3.1*10^8/5*10^3	417.4	226.6
2	Top	6.3	2.1*10^9	8.6*10^7	-	493.7	429.5
	Middle	5.9	5.0*10^7	1.0*10^3	2.0*10^8/5*10^3	297.9	273.5
	Bottom	6.5	8.0*10^7	6.1*10^6	0/1.8*10^4	215.3	204.6
3	Top	6.4	1.0*10^9	8.8*10^7	-	589.9	173.5
	Middle	6.6	4.9*10^6	1.3*10^4	1.2*10^7/5*10^3	589.4	369.9
	Bottom	6.3	4.7*10^6	1.1*10^6	7.4*10^7/5*10^3	258.0	195.3
4	Top	6.8	2.2*10^8	1.0*10^8	-	229.1	251.3
	Middle	6.4	8.9*10^5	8.1*10^3	6.6*10^7/5*10^3	214.0	203.2
	Bottom	6.5	1.2*10^6	8.5*10^3	5.0*10^7/1.8*10^4	290.5	352.2
5	Top	6.5	4.3*10^8	1.0*10^8	-	122.5	349.8
	Middle	5.6	1.7*10^6	7.6*10^4	1.5*10^7/1.8*10^4	313.3	142.9
	Bottom	6.2	1.2*10^6	1.2*10^3	6.6*10^7/5*10^3	294.7	199.2

Note: Not detection.

Chemical analyses have shown that the initial concentration of oil in the models have varied with height of the models from 370 to 550 g/kg DM. The concentration of oil in the models has decreased as per the gravimetric analyses on an average by 24–34 percent, including by 19 percent in the control model after finishing this experiment.

GC analyses of oil products from models as per length are presented in Figures 17.1–17.5.

Results of GC of the analyses show that in control model there is degradation process of oil in the top part. The quantity of peaks practically does not change, but their height and areas decreases (Figure 17.2). The amount of oil products in the control model decreased in the top layer by 54 percent. In the middle of the model the number of peaks decreased by 1 peak, but their area increased. In the bottom of the model the quantity of peaks increased by 1 peak, but the area of peaks and their height (Figure 17.2) significantly increased probably at the expense of an oil filtration down. GC analyses of oil products per length of the model No. 2 are presented in Figure 17.3.

FIGURE 17.1 The laboratory installation, modeling a fenny bog, for development of a new technology of bioremediation of peat, polluted with oil.

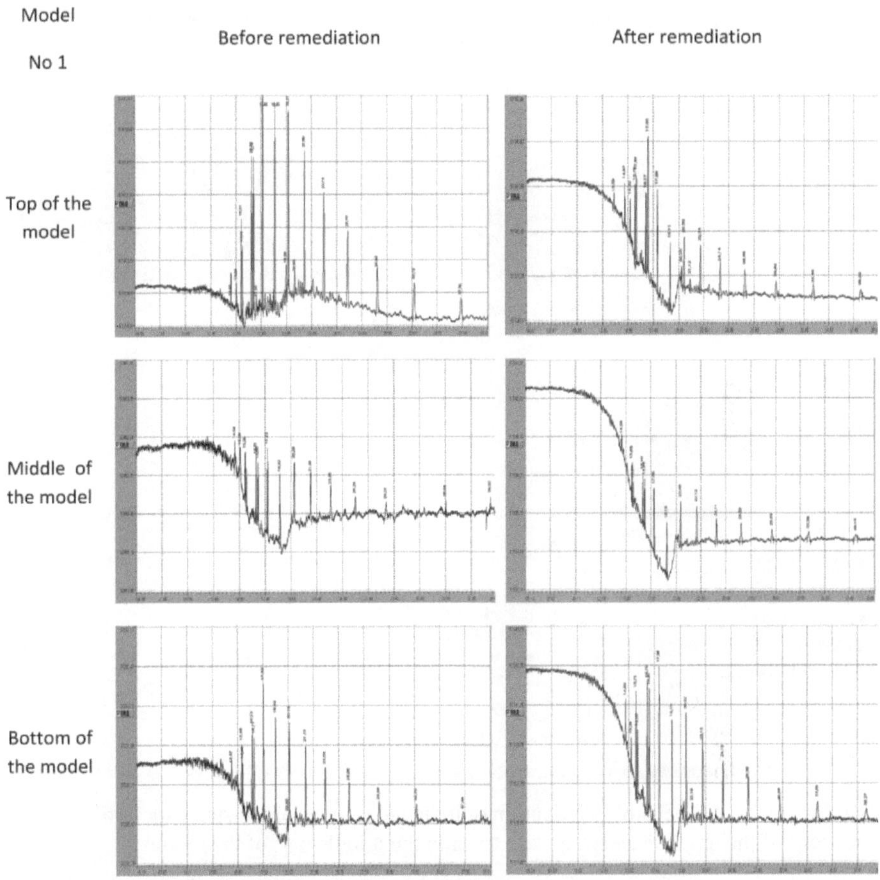

FIGURE 17.2 Model No 1 is a negative control.

Results of GC analyses (Figure 17.3) have shown that processes of oil degradation in the top, middle, and bottom parts of this model have taken place. In the top and the middle parts of this model the quantity of peaks and their areas have decreased. In the bottom part of the model the quantity of peaks does not change, but their areas have slightly decreased. The concentration of oil products in the model decreased in the top part by 74 percent, in the middle part by 24 percent, and in bottom by 5 percent. In this model the gaseous electron acceptor promotes degradation of oil products by anaerobic microorganisms in the middle and bottom parts. But in the top of the model aerobic indigenous microorganisms works.

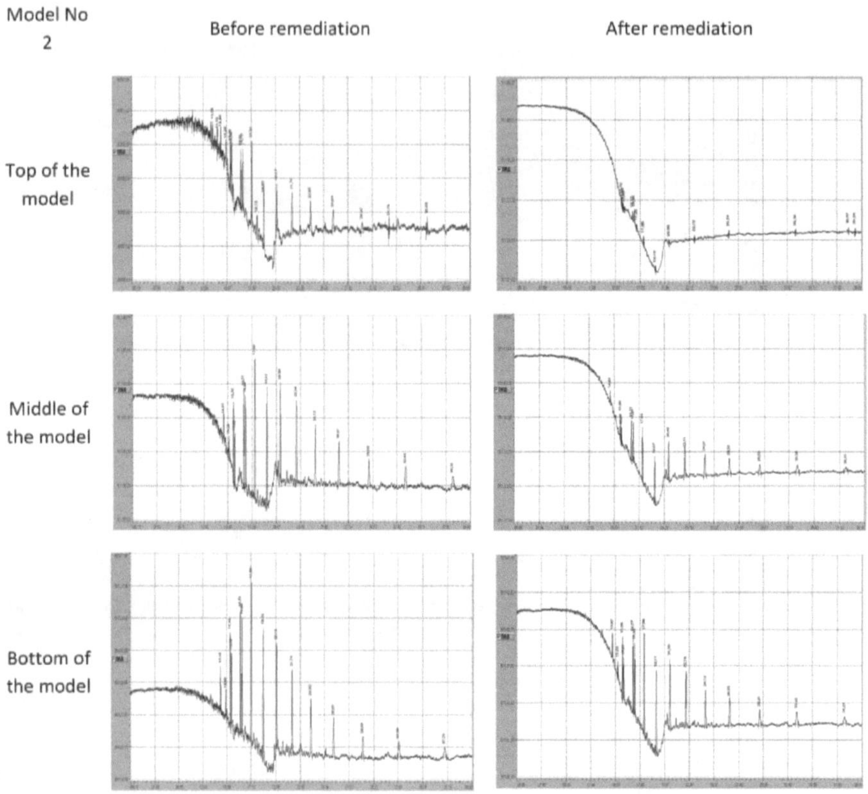

FIGURE 17.3 Model No 2. Activation of indigenous microorganisms with fertilizers and injections of the gaseous electron acceptor into the middle part of the model.

GC analyses of oil products in the model No. 3 is presented in Figure 17.4.

Results of the GC analyses (Figure 17.4) have shown that in the model No 3, in which the Rhoder with the MPN of HCO bacteria 1.0×10^8 cells/mL and fertilizers were added three times along with the gaseous acceptor of electrons injected three times, have activated processes of oil degradation in the top and bottom parts of the model. In the middle part of the model, the quantity of peaks and areas have increased. In the top part of the model the concentration of oil products have decreased by 88 percent, in bottom part by 68 percent. In the middle part of the model, GC analyses have not shown decrease in oil products. The acceptor of electrons probably has promoted degradation of oil products by anaerobic microorganisms in the bottom part of the model.

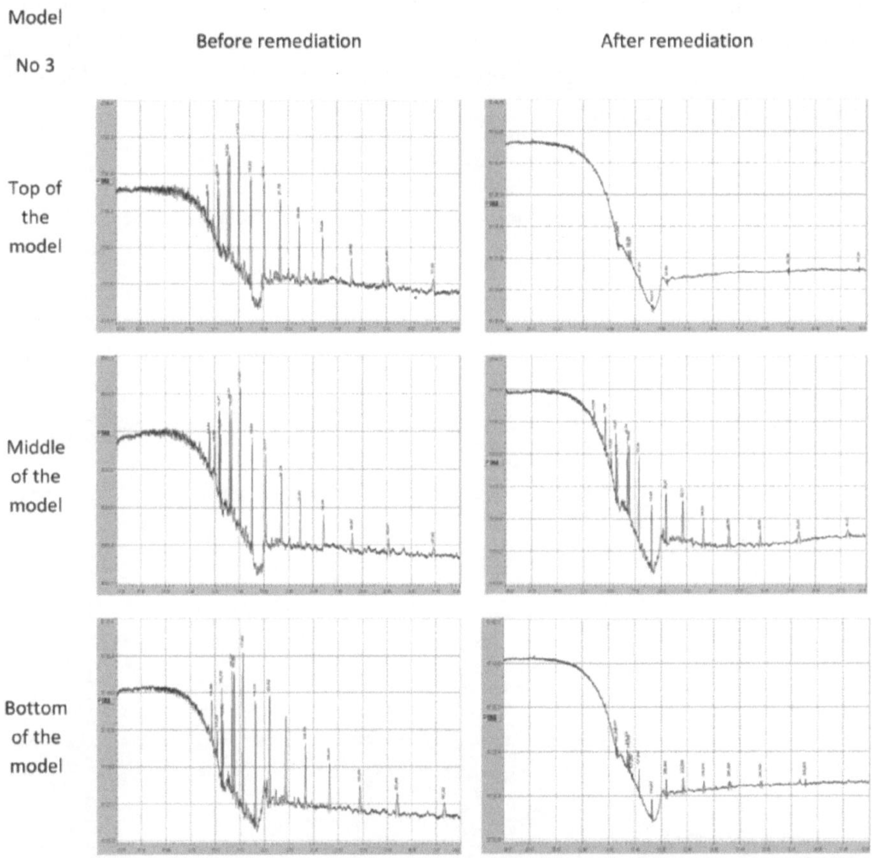

FIGURE 17.4 Model No. 3. Augmentation with the Rhoder and fertilizers, and injection of the gaseous acceptor of electrons into the middle part of the model.

The GC analyses of oil products in the model No. 4 are presented in Figure 17.5.

Results of GC analyses have shown (Figure 17.5) that addition of the Rhoder, fertilizers, and the liquid acceptor of electrons into the top layer of the model three times have decreased the quantity of peaks from 15 to 6 and significantly decreased its areas there. In the middle and the bottom parts of the model the quantity of peaks has not changed, but the areas of peaks in the bottom part of the model have decreased (Figure 17.5). In this model the concentration of oil products have decreased in the top part almost by 96 percent, in the bottom part by 27 percent. In the middle part of the model the areas of peaks have increased by 24 percent by the end of the experiment. Probably the acceptor of electrons added into the top layer of the peat is not so good promoter for degradation of oil products by anaerobic microorganisms.

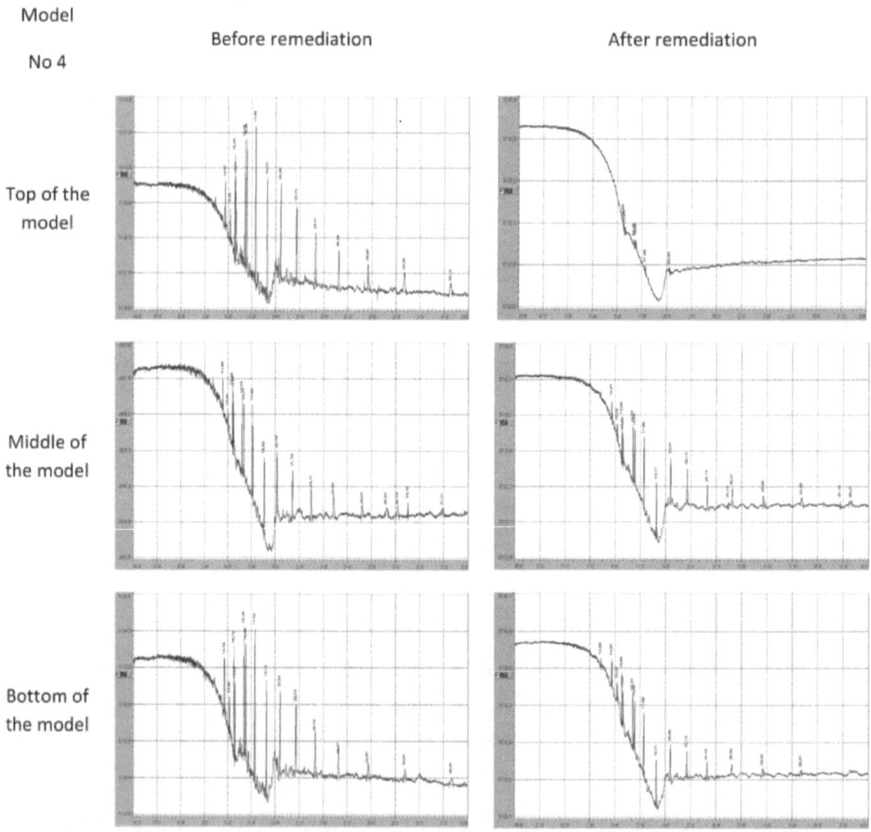

FIGURE 17.5 Model No. 4. Addition of the Rhoder, fertilizers, and the liquid acceptor of electrons into the top layer of the model three times.

The GC analyses of oil products in the model No. 5 are presented in Figure 17.6.

Results of GC analyses have shown that in the top part of the model (Figure 17.6) the quantity of peaks has decreased from 18 to 10 and very significantly its total area has decreased. The concentration of oil products has decreased in the top layer of the peat by 99 percent. In the middle part of the model the quantity of peaks did not change, but the total area of peaks had decreased. And concentration of oil products had decreased on the average by 64 percent. In the bottom part of the model the quantity of peaks had decreased from 19 to 15. The liquid acceptor of electrons, injected into the middle part of the model, significantly promotes degradation of oil products by anaerobic microorganisms in the middle and bottom parts in this model. In the top part of the model the process of oil degradation was provided by Rhoder.

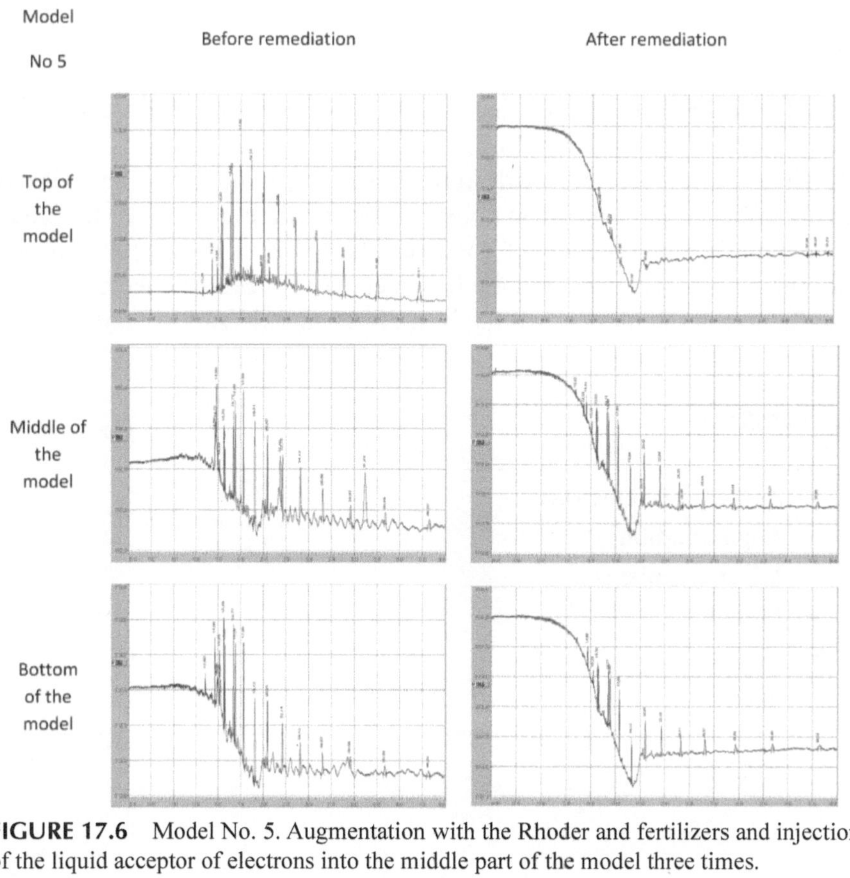

FIGURE 17.6 Model No. 5. Augmentation with the Rhoder and fertilizers and injection of the liquid acceptor of electrons into the middle part of the model three times.

Thus, the received results show that the liquid electron acceptor in comparison with the gaseous electron acceptor has better effect on the degradation of oil, and it is more expedient to inject the liquid electron acceptor into the middle part of the model (into the depth of 40 cm). These results are the first step in development of a new aerobic and anaerobic bioremediation technology for strongly polluted fenny and almost impassable bogs in the North of Russia. Because it is impossible to collect spills of oil completely, it is difficult to perform a classic remediation technology on polluted bogs especially with using specialized equipment and devices. It should be noted that the liquid acceptor of electrons has no relation to iron salts; this acceptor is eco-friendly and activates the indigenous anaerobic microorganisms.

17.4 CONCLUSIONS

The obtained results showed that both studied acceptors of electrons work well in the anaerobic zone. In the aerobic zone, the Rhoder works more effectively in comparison with indigenous microorganisms. The oil oxidizing effect of the Rhoder in combination with the gaseous or the liquid acceptor of electrons showed good results. However augmentation with the Rhoder and fertilizer, and the liquid electron acceptor injected into the middle part of the model, had the best effect on oxidization of oil there.

KEYWORDS

- **Acceptor of electrons**
- **Augmentation**
- **Microorganisms**
- **Model**
- **Oil**
- **Peat**

REFERENCE

1. Murygina, V. P.; Arinbasarov, M. U.; and Kalyuzhnyi, S. V.; Ecology and Industry of Russia. **1999,** *8(16),* (in Russian).
2. Murygina, V.; Arinbasarov, M.; and Kalyuzhnyi, S.; Biodegradation. **2000,** *11(6),* 385, (in Russian).
3. Valentina P. Murygina; Maria Y. Markarova; and Sergey V. Kalyuzhnyi; Environmental International. **2005,** *31(2),* 163.
4. Ouyang, W.; Yu, Y.; Liu, H.; Murygina, V.; Kalyuzhnyi, S.; and Xiu, Z.; Process Biochemistry. **2005,** *40(12),* 3763.
5. Wei Ouyang; Hong Liu; Yong-Yong Yu; Murygina, V.; Kalyuzhnyi, S.; and Zeng-de Xiu; Huanjing Kexue/Environmental Science. **2006,** *27(1),* 160.
6. De-Qing, S.; Jian, Z.; Zhao-Long, G.; Jian, D.; Tian-Li, W.; Murygina, V.; and Kalyuzhnyi, S.; Water, Air, and Soil Pollution. **2007,** *185(1–4),* 177.
7. Valentina Murygina; Maria Markarova; and Sergey Kalyuzhnyi; In Proc. of ipy-osc Symp. Norway, Oslo; **2010,** http://www.ipy-osc.no/
8. Murygina, V.; Gaidamaka, S.; Iankevich, M.; and Tumasyanz, A.; Progress in Environmental Science and Technology. **2011,** III, 791.
9. Drugov, Yu. S.; Zenkevich, I. G.; Rodin, A. A.; Gas chromatography Identification of Air, Water and Soil and Bio-nutrients Pollutants. Binom, Moscow; **2005,** 752 p. (in Russian)
10. Mineev, V. G.; ed. Practical Handbook on Agro Chemistry. Moscow, Russia: Moscow State University; **2001,** 688 p. (in Russian)
11. Netrusov, A. I.; ed. Practical Handbook on Microbiology. Moscow, Russia: Academia; **2005,** 608 p. (in Russian)
12. Nazina, T.; Rozanova, Ye.; Belyayev, S.; and Ivanov, M.; Chemical and Microbiological Research Methods for Reservoir Liquids and Cores of Oil Fields. Pushchino: Preprint Biological Centre Press; **1988,** 35 p. (in Russian)

CHAPTER 18

THE SELECTIVE ETHYLBENZENE OXIDATION CATALYZED WITH THE TRIPLE COMPLEXES NIII(ACAC)2·LIST (OR NAST)·PHOH. ROLE OF H-BONDING AND SUPRAMOLECULAR NANOSTRUCTURES IN MECHANISM OF CATALYSIS

L. I. MATIENKO, L. A. MOSOLOVA, V. I. BINYUKOV, and G. E. ZAIKOV

CONTENTS

18.1 INTRODUCTION

In recent years, the studies in the field of homogeneous catalytic oxidation of hydrocarbons with molecular oxygen were developed in two directions, namely, the free-radical chain oxidation catalyzed by transition metal complexes, and the catalysis by metal complexes that mimic enzymes. Low yields of oxidation products in relation to the consumed hydrocarbon (RH) caused by the fast catalyst deactivation are the main obstacle to the use of the majority of biomimetic systems on the industrial scale [1, 2].

The problem of selective oxidation of alkylarenes to hydroperoxides is economically sound. Hydroperoxides are used as intermediates in large-scale production of important monomers. For instance, propylene oxide and styrene are synthesized from α-phenylethyl hydroperoxide, whereas cumyl hydroperoxide is the precursor in the synthesis of phenol and acetone [1, 2]. The method of modifying the Ni^{II} and $Fe^{II,III}$ complexes used in the selective oxidation of alkylarenes (ethylbenzene and cumene) with molecular oxygen to afford the corresponding hydroperoxides aimed at increasing their selectivity's has been first proposed by L. I. Matienko [3, 4]. This method consists of introducing additional mono- or multidentate modifying ligands into catalytic metal complexes. The mechanism of action of such modifying ligands was elucidated. New efficient catalysts for selective oxidation of ethylbenzene to α-phenylethyl hydroperoxide were developed [3, 4].

Nanostructure science and supramolecular chemistry are fast evolving fields that are concerned with manipulation of materials that have important structural features of nanometer size (1 nm–1 μm) [5, 6]. Nature has been exploiting the noncovalent interactions for the construction of various cell components. For instance, microtubules, ribosomes, mitochondria, and chromosomes use mostly hydrogen bonding in conjunction with covalently formed peptide bonds to form specific structures.

H-bonding can be a remarkably diverse driving force for the self-assembly and self-organization of materials. H-bonds are commonly used for the fabrication of supramolecular assemblies because they are directional and have a wide range of interactions energies. These energies are tunable by adjusting the number of H-bonds, their relative orientation, and their position in the overall structure. H-bonds in the center of protein helices can have energy of 20 kcal/mol because of their cooperative dipolar interactions [7].

The porphyrin linkage through H-bonds is the binding type that is generally observed in nature. One of the simplest artificial self-assembling supramolecular porphyrin systems is the formation of a dimer, based on carboxylic acid functionality [8].

The mechanism of catalysis often involves the formation of a supramolecular assembly during the reaction [8].

In the present article we research the phenomenon of unusually high efficiency of triple catalytic systems {$Ni^{II}(acac)_2$+L^2+PhOH} (L^2=MP, MP=N-metylpirrolidon-2), {$Ni^{II}(acac)_2$+NaSt(or LiSt)+PhOH} in the selective ethylbenzene oxidation by dioxygen into α-phenylethyl hydroperoxide, with which we began [3, 4]. The features of the mechanism of triple complexes catalysis are the stability of complexes $Ni^{II}(acac)_2$·NaSt(or LiSt)·PhOH, the H-bonding, and the supramolecular structures formation with assistance of intermolecular H-bonds during ethylbenzene oxidation. Earlier we have proposed a new approach with the AFM method for researching the possibility of supramolecular structure formation because of H-bonds based on catalytic active nickel complexes [9], including $Ni^{II}(acac)_2$·NaSt·PhOH [10]. Here the AFM method was applied also for research of possibility of the formation of stable supramolecular nanostructures based on the bimetallic, heteroligand complexes $Ni^{II}(acac)_2$·LiSt·PhOH because of intermolecular H-bonds.

18.2 EXPERIMENT

Ethylbenzene (RH) was oxidized with dioxygen at 120°C in glass bubbling-type reactor in the presence of three-component systems {$Ni^{II}(acac)_2$+L^2+PhOH} (L^2=NaSt, LiSt)) [3, 4].

Analysis of oxidation products. α-Phenylethyl hydroperoxide (PEH) was analyzed by iodometry. By-products, including methylphenylcarbinol (MPC), acetophenone (AP), and phenol (PhOH) as well as the RH content in the oxidation process were examined by GLC [3, 4].

The order in which PEH, AP, and MPC formed was determined from the time dependence to product accumulation rate ratio at $t \rightarrow 0$. The variation of these ratios with time was evaluated by graphic differentiation [3, 4].

Experimental data processing was done using special computer programs Mathcad and Graph2Digit.

AFM SOLVER P47/SMENA/ with Silicon Cantilevers NSG11S (NT MDT) with curvature radius 10 nm, tip height 10–15 μm, and cone angle ≤22° in taping mode on resonant frequency 150 kHz was used.

As substrate, the polished Silicone surface special, chemically modified was used.

Modified Silicone surface was exploited for its self-assembly driven growth because of H-bonding of complexes $Ni^{II}(acac)_2$·LiSt·PhOH with Silicone surface. The saturated solution of complexes was put on a surface, maintained for some time, and then solvent was deleted from the surface by means of special method–spin-coating process.

In the course of scanning of investigated samples, it has been found, that the structures are fixed on the surface strongly enough because of H-bonding. The self-assembly driven growth of the supramolecular structures on the basis of complexes, mentioned above are because of H-bonds and perhaps the other noncovalent interactions observed on Silicone surface.

18.3 RESULTS AND DISCUSSION

Mechanism of catalysis with triple systems {$Ni(II)(acac)_2$+L^2+PhOH}

Role of intramolecular H-bonding

The phenomenon of a substantial increase in the selectivity (***S***) and conversion (***C***) of the ethylbenzene oxidation to the to α-phenylethyl hydroperoxide upon addition of PhOH together with alkali metal stearates MSt (M=Li, Na) as metalloligands to metal complexes $Ni^{II}(acac)_2$ was discovered in our works [3, 4].

The observed values of ***C*** [$C > 35\%$ at $(S_{PEH})_{max} \sim 90\%$], $[ROOH]_{max}$ (1.6−1.8 mol L^{-1})] far exceeded those obtained with other ternary catalytic systems {$Ni^{II}(acac)_2$+L^2+PhOH} (L^2 is N-metylpyrrolidone-2 (MP), hexamethylphosphorotriamide (HMPA),) and the majority of active binary systems (Figure 18.1). These results by L. I. Matienko and L. A. Mosolova are protected by the Russian Federation patent (2004).

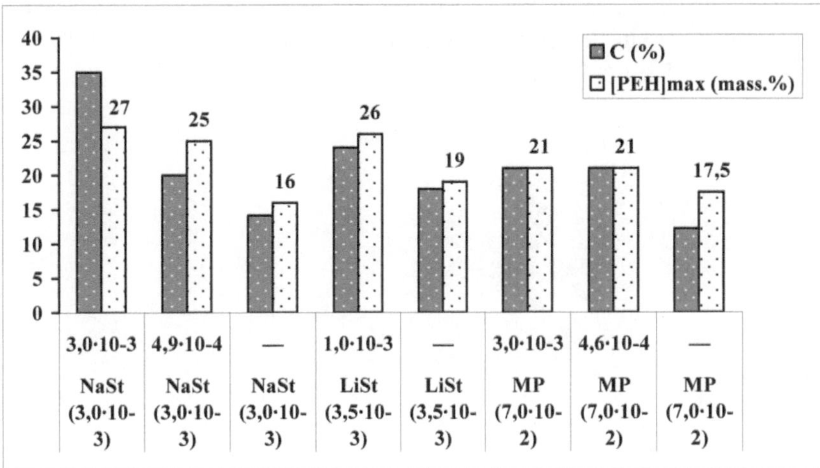

FIGURE 18.1 Values of conversion ***C***(%) (I row), maximum values of hydroperoxide concentrations $[PEH]_{max}$ (mass.%) (II row) in reactions of ethylbenzene oxidation in the presence of triple catalytic systems {$Ni(II)(acac)_2+L^2+PhOH$} (L^2=NaSt, LiSt, MP: [PhOH] (mol/L)—on an axis of abscises (the top number), [L^2] (mol/L)—on an axis of abscises (the bottom number)). $[Ni^{II}(acac)_2]=3.0\cdot10^{-3}$ mol/L, 120°C.

There are characteristic features for triple systems including metallo-ligand–modifier L^2=NaSt, LiSt (and N-metylpirrolidon-2) compared with the most active binary systems (Figure 18.2). The advantage of these ternary systems is the long-term activity of the *in* situ formed complexes $Ni^{II}(acac)_2\cdot L^2\cdot PhOH$. Unlike binary systems, the $acac^-$ligand in nickel complex does not undergo transformations in the course of ethylbenzene oxidation as in this case. (The formation of triple complexes $Ni^{II}(acac)_2\cdot L^2\cdot PhOH$ at very early stages of oxidation was established with kinetic methods [3, 4]). So the reaction rate remains practically the same during the oxidation process. In the course of the oxidation the rates of products accumulation remained unchanged during the long period t ≤ 30–40 h (as one can see on Figure 18.2a, b).

At catalysis with {$Ni^{II}(acac)_2$+MSt+PhOH} the oxidation products PEH, AP, MPC are formed practically without the autoacceleration period (NaSt) or with maximum initial rate (LiSt) (unlike catalysis with triple system which does not contain metalloligand (L^2=MP)). Throughout the process of ethylbenzene oxidation, the rates of formation of PEH (and

AP, MPC also): w_P=constant=maximum (P=PEH, AP, or MPC) (NaSt); w_P=constant=maximum (P=AP, MPC), w_P=constant ($<w_{max}$) (P=PEH) in the case of L^2=LiSt (Figure 18.2).

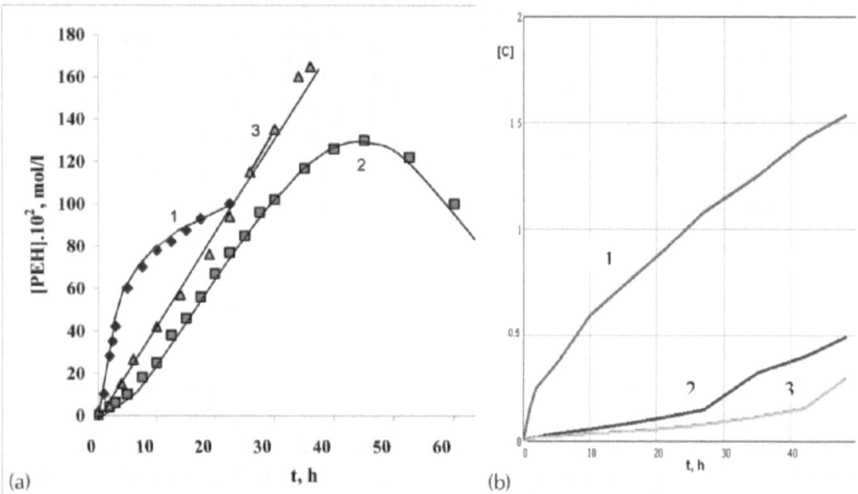

FIGURE 18.2 (a) Kinetics of accumulation of PEH in reactions of ethylbenzene oxidation, catalyzed by binary system $\{Ni^{II}(acac)_2+MP\}$ (1) and two triple systems $\{Ni^{II}(acac)_2+L^2+PhOH\}$ with L^2=MP (2) and L^2=NaSt (3). $[Ni^{II}(acac)_2]=3\cdot10^{-3}$ mol/L, and $[MP]=7\cdot10^{-2}$ mol/L, $[NaSt]=3\cdot10^{-3}$ mol/L, $[PhOH]=3\cdot10^{-3}$ mol/L. 120°C and (b) Kinetics of accumulation of PEH (1), AP (2), MPC (3) in ethylbenzene oxidation, catalyzed by triple system $\{Ni^{II}(acac)_2+LiSt+PhOH\}$. $[Ni^{II}(acac)_2]=3\cdot10^{-3}$ mol/L, $[LiSt]=3\cdot10^{-3}$ mol/L, $[PhOH]=3\cdot10^{-3}$ mol/L. 120°C (Data in Figure 18.2(b) are presented for the first time).

Often metals of constant valency compounds are used in combination with redox-active, transition-metal complexes to promote a variety of reactions involving the transfer of electrons [11]. This effect is typified in metalloproteins such as the copper, zinc, superoxide dismutase, in which both metal ions have been proposed to be functionally active [11].

Earlier we have established, that the increase in the initial rate of the ethylbenzene oxidation with dioxygen, catalyzed with $Ni^{II}(acac)_2$ in the presence of additives of metalloligands MSt (M=Li, Na, K), is because of higher activity of formed complexes $Ni^{II}(acac)_2\cdot MSt$ in the microstages of chain initiation (O_2 activation) and/or decomposition of PEH with free radical formation [3, 4].

The participation of catalyst $Ni^{II}(acac)_2$•MSt in microsteps of chain propagation, and probably in chain termination must also be taken into account [3, 4]. The results found in [11] illustrate the possibility that redox-inactive metal ions can be used to facilitate the activation of dioxygen, and this is consistent with our data.

At catalysis with triple complexes $Ni(acac)_2$•L^2•PhOH (L^2=NaSt, LiSt) the parallel formation of α-phenylethyl hydroperoxide (PEH), acetophenone (AP), and methylphenylcarbinol (MPC) was observed ($w_{AP(MPC)}/w_{PEH}$ 0 a t t ® 0, w_{AP}/w_{MPC} 0 a t t ® 0) throughout the reaction of ethylbenzene oxidation) (see, for example, Figures 18.3a, b).

A more considerable increase in the selectivity (S_{PEH}) at the catalysis by $Ni^{II}(acac)_2$•L^2•PhOH (L^2=NaSt, LiSt) complexes are seen compared with noncatalytic oxidation, wherein the binary systems {$Ni^{II}(acac)_2$+MSt} was associated with the change in the route of acetophenone and methylphenylcarbinol formation: AP and MPC form in parallel with PEH rather than as a result of PEH decomposition, and with the inhibition of the PEH heterolytic decomposition with PhOH formation.

FIGURE 18.3 (a) Dependence $\Delta[AP]_{ij}/\Delta[PEH]_{ij}\cdot10^2$ on time t_j in the course of ethylbenzene oxidation, catalyzed with complexes $Ni^{II}(acac)_2$·NaSt·PhOH (1:1:1), 120°C and (b) Dependence $\Delta[AP]_{ij}/\Delta[PEH]_{ij}\cdot10^2$ on time t_j in the course of ethylbenzene oxidation, catalyzed with complexes $Ni^{II}(acac)_2$·LiSt·PhOH (1:1:1), 120°C.

Thus we had shown that the triple complexes $Ni^{II}(acac)_2$·NaSt(LiSt)·PhOH unlike binary complexes $Ni^{II}(acac)_2$·NaSt(LiSt) obviously were inactive in the reaction of hydroperoxide decomposition. However, the ability of redox-inactive metal ions to facilitate the activation of dioxygen (and free radical formation in chain initiation) continued in the case of the catalysis with the triple complexes, $Ni^{II}(acac)_2$·NaSt(LiSt)·PhOH. So catalysis with

$Ni^{II}(acac)_2 \cdot NaSt(LiSt) \cdot PhOH$ is largely associated with the triple complexes involvement in the steps of chain initiation (activation of O_2) and chain propagation ($Cat + RO_2^{\bullet} \rightarrow$).

In these systems, dioxygen activation may be promoted through the formation of intramolecular H-bonds [3, 4]. As can be seen from Figure 18.2, complexes $Ni^{II}(acac)_2 \cdot LiSt \cdot PhOH$, probably, have a higher activity in relation to molecular oxygen compared to complexes $Ni^{II}(acac)_2 \cdot NaSt(MP) \cdot PhOH$.

The role of intramolecular H-bonds in the catalysis mechanism was established by us in mechanism of formation of triple catalytic complexes $\{Ni^{II}(acac)_2 \cdot L^2 \cdot PhOH\}$ (L^2=N-methylpirrolidon-2) in the process of ethylbenzene oxidation with molecular oxygen [3, 4].

We have established that concentration [PhOH] at catalysis with system $\{Ni^{II}(acac)_2$ ($3.0 \cdot 10^{-3}$ mol/l)+NaSt(LiSt) ($3.0 \cdot 10^{-3}$ mol/L)+PhOH ($3.0 \cdot 10^{-3}$ mol/L)$\}$, as well as at catalysis with similar system $\{Ni^{II}(acac)_2$ ($3.0 \cdot 10^{-3}$ mol/L)+MP ($7.0 \cdot 10^{-2}$ mol/L)+PhOH ($3.0 \cdot 10^{-3}$ mol/L)$\}$ including MP as donor exoligand L^2, decreases during the first hours of oxidation (Figures 18.4a, 18.5a). These changes in PhOH concentrations seem to be because of the triple complexes $Ni^{II}(acac)_2 \cdot L^2 \cdot PhOH$ formation [3,4].

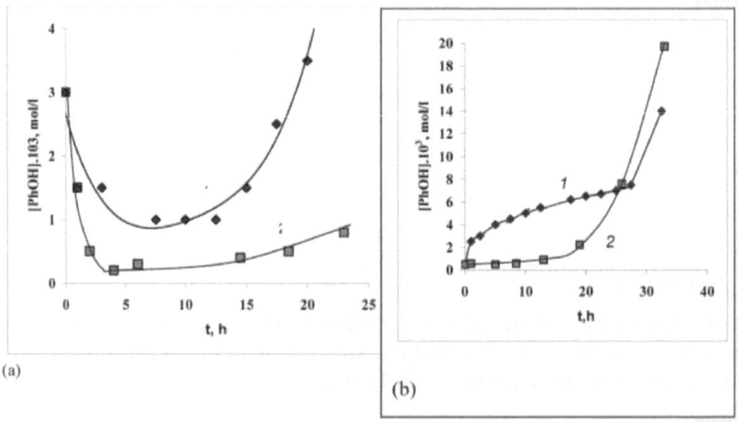

FIGURE 18.4 **(A)** Kinetics of accumulation of PhOH in reaction of ethylbenzene oxidation catalyzed by and two triple systems $\{Ni^{II}(acac)_2$+MP+PhOH$\}$ (*1*,◊) and $\{Ni(II)(acac)_2$+NaSt+PhOH$\}$ (*2*, □) with [PhOH] = $3 \cdot 10^{-3}$ mol/L; (b) Kinetics of accumulation of PhOH in reaction of ethylbenzene oxidation catalyzed by two triple systems $\{Ni^{II}(acac)_2$+MP+PhOH$\}$ (1) and $\{Ni^{II}(acac)_2$+NaSt+PhOH$\}$ (2) with smaller concentration [PhOH]=$4.6 \cdot 10^{-4}$ mol/L ($4.9 \cdot 10^{-4}$ mol/l). $[Ni^{II}(acac)_2]=3 \cdot 10^{-3}$ mol/L, [MP]=$7 \cdot 10^{-2}$ mol/L, [NaSt]=$3 \cdot 10^{-3}$ mol/L, 120°C.

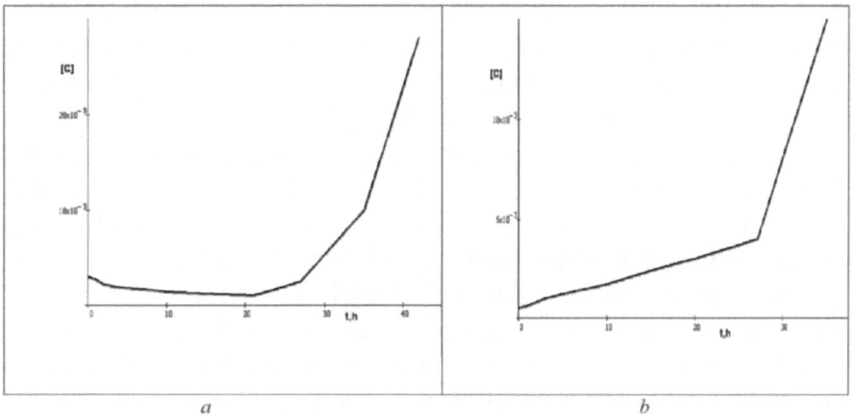

FIGURE 18.5 Kinetics of accumulation of PhOH in reaction of ethylbenzene oxidation catalyzed by the triple system {$Ni^{II}(acac)_2$ ($3.0 \cdot 10^{-3}$ mol/L)+LiSt ($3.0 \cdot 10^{-3}$ mol/L)+PhOH} with different initial concentrations of PhOH: $[PhOH]_0 = 3 \cdot 10^{-3}$ mol/L (*a*) or $[PhOH]_0 = 5 \cdot 10^{-4}$ mol/L(*b*). These data are received at the first time.

In the presence of triple systems with smaller concentrations of phenol {$Ni^{II}(acac)_2$+L^2+PhOH (~$5 \cdot 10^{-4}$ mol/L)} (Figure 18.4b, 18.5b) catalysis of ethylbenzene oxidation seems to be because of activity of complexes $[(M(L^1)_2)_m \cdot (L^2)_n \cdot (PhOH)_q]$, formed on initial steps of reaction.

At catalysis by triple system {$Ni^{II}(acac)_2$+MP+PhOH} with small phenol concentration [PhOH]=$4.6 \cdot 10^{-4}$ mol/L the fast increase but not decrease in the concentration of PhOH is observed right up to [PhOH]=$(3-5) \cdot 10^{-3}$ mol/L (Figure 18.4(b), curve 1). [PhOH] = $(3–5) \cdot 10^{-3}$ mol/L corresponds to concentration [PhOH] for the first combination {$Ni^{II}(acac)_2$ ($3.0 \cdot 10^{-3}$ mol/L) + MP ($7.0 \cdot 10^{-2}$ mol/L)+PhOH ($3.0 \cdot 10^{-3}$ mol/L)}, and apparently to the formation of complexes of structure $[M(L^1)_2 \cdot (L^2) \cdot (PhOH)]$ [3, 4].

The rate of PhOH formation at the beginning of the ethylbenzene oxidation in the presence of {$Ni^{II}(acac)_2$+NaSt+PhOH} with small phenol concentration [PhOH] of $5 \cdot 10^{-4}$ mol/L (Figure 18.4(b), curve 2) or {$Ni^{II}(acac)_2$+LiSt+PhOH} (Figure 18.5(b) is much less compared to the catalysis with {$Ni^{II}(acac)_2$+MP+PhOH}.

The increase in the concentration of PhOH (a result of PEH heterolysis) at the beginning of the process may be because of the function of PhOH as an acid that became stronger as a consequence of outer sphere coordination of PhOH with nickel complex $Ni^{II}(acac)_2 \cdot L^2$ [3, 4]. This presumption is confirmed by the next facts [3, 4]. So the accumulation of PhOH (but not the consumption) with maximum initial rate $w_{PhOH,0}=w_{PhOH,max}$ is observed upon addition of PhOH ($3.0 \cdot 10^{-3}$ mol/L) into the reaction of ethylbenzene oxidation catalyzed by coordinated saturated complexes $Ni^{II}(acac)_2$ 2MP ($[Ni^{II}(acac)_2] = 3.0 \cdot 10^{-3}$ mol/L, $[MP] = 2.1 \cdot 10^{-1}$ mol/L) (Figure 18.6), and also in the case of the ethylbenzene oxidation catalyzed by binary system {$Ni^{II}(acac)_2$ ($3.0 \cdot 10^{-3}$ mol/L)+PhOH($5 \cdot 10^{-4}$ mol/L)} at [MP] = 0.

Outer sphere coordination of phenol to a complex of nickel $Ni^{II}(acac)_2 \cdot 2MP$ does not influence on catalytic activity of the last–on the order in which the oxidation products form. We have established that upon addition of PhOH ($3.0 \cdot 10^{-3}$ mol/l) into the reaction of ethylbenzene oxidation catalyzed by coordinated saturated complexes $Ni^{II}(acac)_2 \cdot 2MP$ ($[Ni^{II}(acac)_2]=3.0 \cdot 10^{-3}$ mol/L, $[MP] = 2.1 \cdot 10^{-1}$ mol/L), namely at outer sphere coordination of PhOH, which obviously takes place under these conditions, parallel formation of major products of ethylbenzene oxidation as well as in the case of $Ni^{II}(acac)_2 \cdot 2MP$ was observed. So in the ethylbenzene oxidation, catalyzed with $Ni^{II}(acac)_2 \cdot 2MP \cdot PhOH$ the oxidation products PEH, AP, and MPC are formed parallel, AP and MPC—also parallel each other: $w_P/w_{PEH} \neq 0$ at $t \to 0$, and $w_{AP}/w_{MPC} \neq 0$ at $t \to 0$ (P=AP or MPC). At that the value of S_{PEH} is maximal on initial stages, and $S_{PEH,0} \leq 90\%$ at conversion $C \sim 10\%$ (Figure 18.6).

At the ethylbenzene oxidation catalyzed by binary system { $Ni^{II}(acac)_2$ ($3.0 \cdot 10^{-3}$ mol/L)+PhOH($5 \cdot 10^{-4}$ mol/L)} at [MP] = 0 at the beginning stages of reaction the side products AP, and MPC are formed successively, at decomposition of PEH, and AP and MPC—also successively, namely, $w_P/w_{PEH} \to 0$ at $t \to 0$ and $w_{AP}/w_{MPC} \to 0$ at $t \to 0$ (P=AP, or MPC), as at catalysis with $Ni^{II}(acac)_2$ ($3.0 \cdot 10^{-3}$ mol/L) only [4]. Further, in process of development of oxidation reaction, the route of formation of products changes apparently owing to outer sphere coordination of PhOH to $Ni^{II}(acac)_2$: $w_P/w_{PEH} \neq 0$ at $t \to 0$, and $w_{AP}/w_{MPC} \neq 0$ at $t \to 0$ (P=AP or MPC) (as one can see on Figure 18.7).

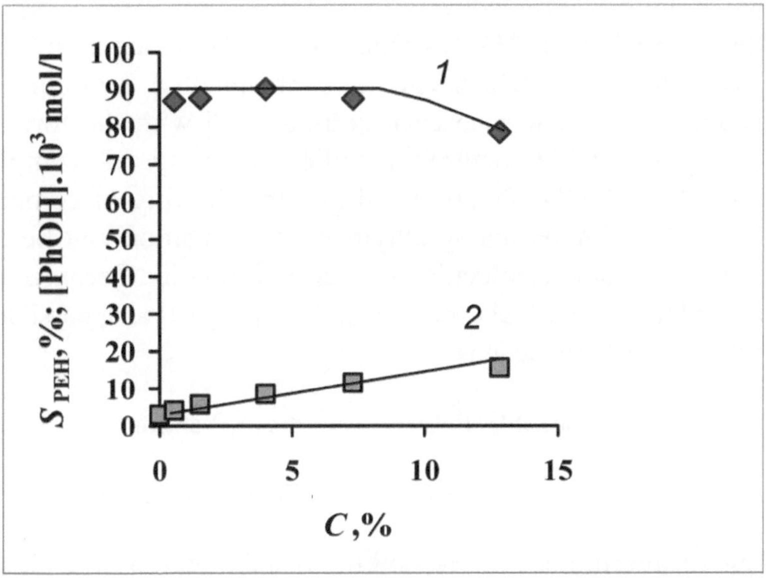

FIGURE 18.6 The dependences of S_{PEH} (1) and [PhOH] (*2*) от ***C*** in reaction of ethylbenzene oxidation in the presence of triple system {$Ni^{II}(acac)_2$ ($3.0 \cdot 10^{-3}$mol/L)+MP($2,1 \cdot 10^{-1}$ mol/L)+PhOH ($3.0 \cdot 10^{-3}$ mol/L)}, 120°C.

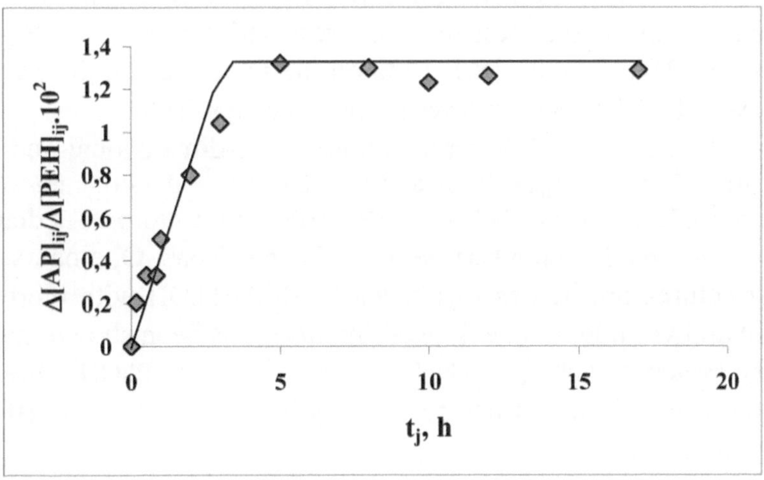

FIGURE 18.7 Dependence $\Delta[AP]_{ij}/\Delta[PEH]_{ij} \cdot 10^2$ on time t_j in the course of ethylbenzene oxidation, catalyzed with system {$Ni^{II}(acac)_2$+PhOH}. [$Ni^{II}(acac)_2$] = $3 \cdot 10^{-3}$ mol/L, [PhOH] =$5 \cdot 10^{-4}$ mol/L, 120°C.

The high efficiency of three-component systems {$Ni^{II}(acac)_2$+MSt+PhOH} (M=Na, Li) in the reaction of selective oxidation of ethylbenzene to α-phenylethyl hydroperoxide on parameters *S*, *C*, *w* = constant is associated with the formation of extremely stable heterobimetallic, heteroligand complexes $Ni(acac)_2$•MSt•PhOH. We assumed that the stability of complexes $Ni^{II}(acac)_2$•MSt•PhOH during ethylbenzene oxidation can be associated with the supramolecular structures formation because of intermolecular H-bonds (phenol–carboxylate) [12–14] and, possible, the other noncovalent interactions:

$$\{Ni^{II}(acac)_2+NaSt(\text{or } LiSt)+PhOH\} \rightarrow Ni^{II}(acac)_2\cdot NaSt(\text{or } LiSt)\cdot PhOH \rightarrow$$

$$\rightarrow \{Ni^{II}(acac)_2\cdot NaSt(\text{or } LiSt)\cdot PhOH\}_n$$

Role of intermolecular H-bonding in stabilization of triple catalytic complexes $Ni^{II}(acac)_2 \cdot L^2 \cdot PhOH$.

The association of triple complexes $Ni^{II}(acac)_2$·NaSt(or LiSt)·PhOH to supramolecular structures because of intermolecular H-bonding should be followed from analysis of AFM data, which we received in our works. Results are presented on the next Figures (18.8–18.13) and Table 18.1. Data on the structures on the basis of $Ni^{II}(acac)_2$·LiSt·PhOH complexes are presented for the first time.

Figures 18.8, 18.9 demonstrated three-dimensional and two-dimensional AFM image (30 30 and 10 10 (μm)) of the structures on the basis of triple complexes $Ni^{II}(acac)_2$·NaSt·PhOH [10] formed at drawing of a uterine solution on a surface of modified silicone. One can watch the structures on the basis of $Ni^{II}(acac)_2$·NaSt·PhOH with increased height and volume. In check experiments it has been shown that for binary systems {$Ni^{II}(acac)_2$+NaSt}, and {$Ni^{II}(acac)_2$+PhOH}, the formation of the similar structures (exceeding the height of 2–10 nm) is not observed.

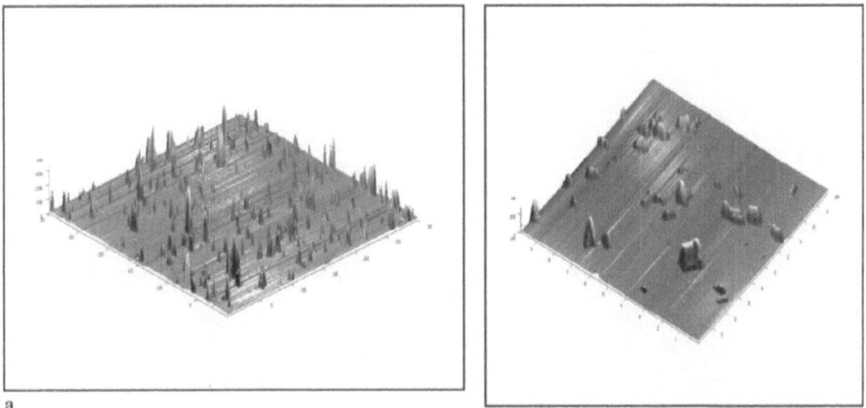

a b

FIGURE 18.8 The AFM three-dimensional image (30 30 (a) and 10 10 (b) (μm)) of the structures formed on a surface of modified silicone on the basis of triple complexes $Ni^{II}(acac)_2 \cdot NaSt \cdot PhOH$.

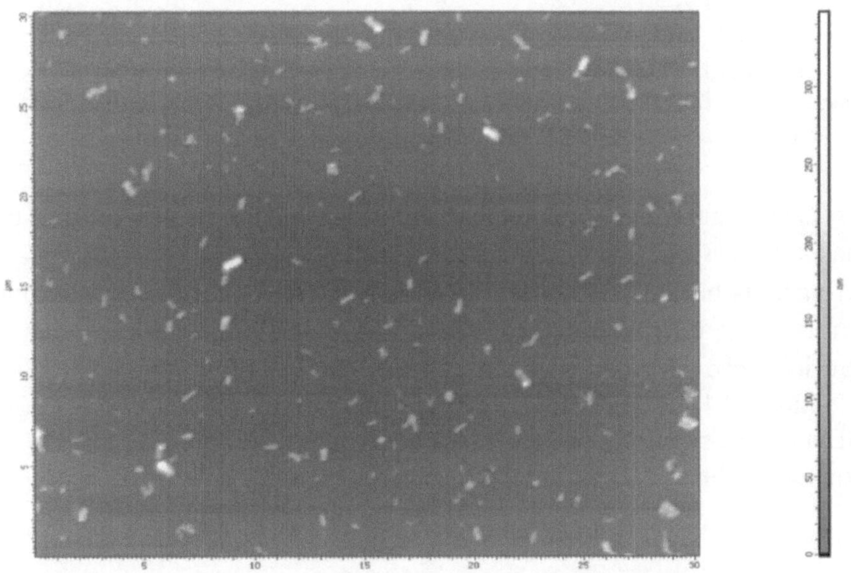

FIGURE 18.9 The AFM two-dimensional image (30 ′ 30 (μm)) of nanoparticles on the basis $Ni^{II}(acac)_2 \cdot NaSt \cdot PhOH$ formed on the surface of modified silicone.

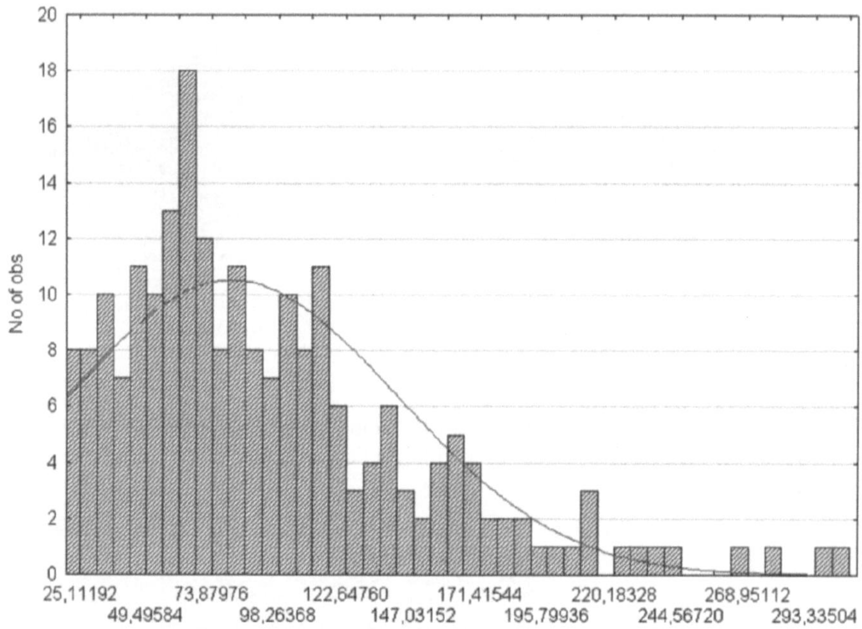

FIGURE 18.10 Histogram of mean values of height (nm) of the AFM images of nanostructures based on $Ni^{II}(acac)_2 \cdot NaSt \cdot PhOH$, formed on the surface of modified silicone.

In the Figure 18.10, a histogram with mean height of nanoparticles on basis of $Ni^{II}(acac)_2 \cdot NaSt \cdot PhOH$ is presented. As can see, structures are of various heights from the 25 nm to the 250–300 nm for maximal values. The distribution histogram shows that the greatest number of particles are particles with a mean size of 50–100 nm in height.

Table 18.1 shows the mean values of area, volume, height, width, length of nanoscale structures on basis of triple complexes $Ni^{II}(acac)_2 \cdot NaSt \cdot PhOH$ formed on the surface of modified silicone

TABLE 18.1 The mean values of area, volume, height, length, width of the AFM image of nanoparticles on the basis of $Ni^{II}(acac)_2$·NaSt·PhOH formed on the surface of modified silicone.

Variable	Mean Values	Confidence −95.000%	Confidence +95.000%
Area (μm^2)	0.13211	0.11489	0.14933
Volume (μm^3)	14.11354	11.60499	16.62210
Z (Height) (nm)	80.56714	73.23940	87.89489
Length (μm)	0.58154	0.53758	0.62549
Width (μm)	0.19047	0.17987	0.20107

On the next pictures (Figure 18.11–18.13) the nanoparticles image on the basis of $Ni^{II}(acac)_2$·LiSt·PhOH and $Ni^{II}(acac)_2$·LiSt complexes is demonstrated.

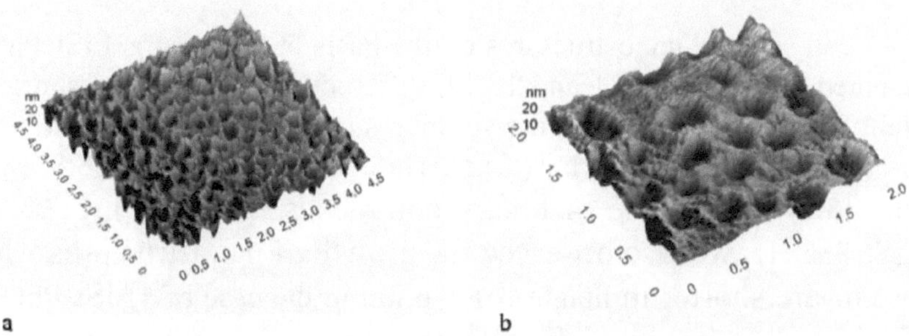

FIGURE 18.11 The AFM three-dimensional image (4.5 × 4.5 (a) and 2× 2 (b) (μm)) of the structures formed on a surface of modified silicone on the basis of triple complexes $Ni^{II}(acac)_2$·LiSt·PhOH.

As one can see, nanostructures on the basis of triple complexes $Ni^{II}(acac)_2$·LiSt·PhOH have the interesting cell form with cell height of 10 nm and cell width of 0.5 μm.

On the Figure 18.12 three- and two-dimensional image (5 × 5 and 2 × 2 (μm)) and profile of one of the nanostructures on the basis of triple complexes $Ni^{II}(acac)_2 \cdot LiSt \cdot PhOH$ with more simple form (cell height of 7–12 nm and cell width of 60 nm), which we observed also on the surface of modified silicone, are presented.

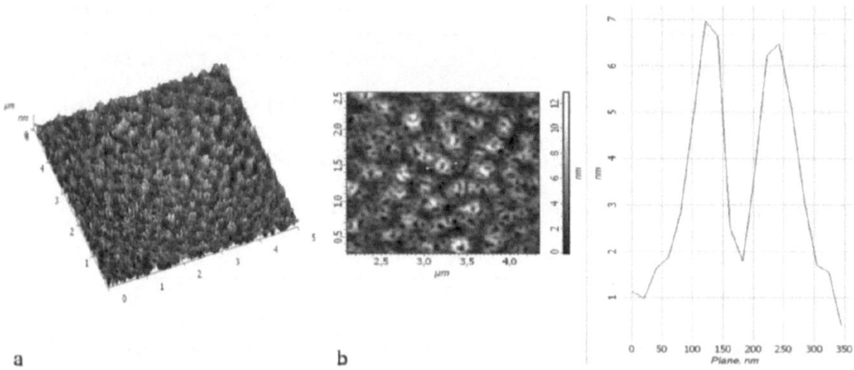

FIGURE 18.12 The AFM three- and two-dimensional image (5 × 5 (a) and 2.5 × 2.5 (b) (μm)) of the structures with more simple form received on a surface of modified silicone on the basis of triple complexes $Ni^{II}(acac)_2 \cdot LiSt \cdot PhOH$, and profile of one of these structures (c).

As can see, the nanostructures on the basis of $Ni^{II}(acac)_2 \cdot LiSt \cdot PhOH$, presented on Figure 18.11 and 18.12, different in form and are shorter in height (h of 10–12 nm) than the structures of complexes which are sodium-based (h ~ 80 nm for $\{Ni^{II}(acac)_2 \cdot NaSt \cdot PhOH\}_n$. (Figure 18.8–18.10, Table 18.1) In the case of binary complexes, $\{Ni^{II}(acac)_2 \cdot LiSt\}$ and $\{LiSt \cdot PhOH\}$, we also observed the growth of nanostructures which were or are shorter in height (of 4–6 nm in the case of $\{LiSt \cdot PhOH\}$) or had less clearly expressed regular structure ($\{Ni^{II}(acac)_2 \cdot LiSt\}$ (see, for example, Figure 18.13), than nanostructures on the basis of triple complexes $\{Ni^{II}(acac)_2 \cdot LiSt \cdot PhOH\}$.

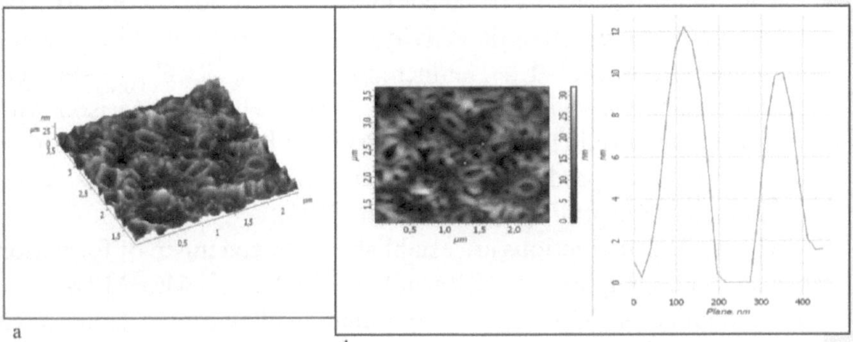

a b

FIGURE 18.13 The AFM three-dimensional image (2.5 × 2.5 (μm)) (a), two-dimensional image (2.5 × 2.5 (μm)) and profile (b) of the nanostructures formed on a surface of modified silicone on the basis of $Ni^{II}(acac)_2 \cdot LiSt$ complexes.

So, we can conclude that the high degrees of conversion of ethylbenzene to PEH and the yields of α-phenylethyl hydroperoxide in the case of catalysis of three-component systems $\{Ni^{II}(acac)_2 + MSt + PhOH\}$ (M=Na, Li) in the reaction of selective oxidation of ethylbenzene to α-phenylethyl hydroperoxide, may be associated as follows: The formation of heterobimetallic, heteroligand complexes $Ni^{II}(acac)_2 \cdot MSt \cdot PhOH$, which are self-organized during ethylbenzene oxidation to form extremely stable supramolecular structures $\{Ni^{II}(acac)_2 \cdot NaSt(\text{or } LiSt) \cdot PhOH\}_n$ at expense of intermolecular (phenol–carboxylate) H-bonds [12–14] and, possibly the other non-covalent interactions. The higher efficiency of heterobinuclear, heteroligand complexes $Ni^{II}(acac)_2 \cdot NaSt \cdot PhOH$, including metalloligand NaSt, as selective catalysts compared with $Ni^{II}(acac)_2 \cdot LiSt \cdot PhOH$, seems to be because of formation of more stable supramolecular structures $\{Ni^{II}(acac)_2 \cdot NaSt \cdot PhOH\}_n$. The data in Figures 18.8–18.13 show in favor of that assumption.

18.4 CONCLUSION

(1) It has been established that inclusion of phenol in coordination sphere of a complex $Ni(II)(acac)_2 \cdot L^2$ (L^2=MSt(M=Na, Li), MP)

leads to formation of a triple complex $Ni(II)(acac)_2 \cdot L^2 \cdot PhOH$ with essentially other catalytic activity, as compared with binary complexes $Ni(II)(acac)_2 \cdot L^2$. The increase in selectivity $\boldsymbol{S}_{\mathbf{PEH,ma}}$~90 percent at catalysis with $Ni(II)(acac)_2 \cdot L^2 \cdot PhOH$ in comparison with noncatalyzed oxidation ($\boldsymbol{S}_{\mathbf{PEH,max}}$=80%) is because of change of a order in which products PEH, AP, and MPC forms (successive via PEH decomposition → parallel with PEH).

The H-bonding interactions are established in mechanism of formation of triple catalytic complexes $Ni(II)(acac)_2 \cdot L^2 \cdot PhOH$ (L^2=MSt(MP)).

(2) We applied AFM method in the analytical purposes to research the possibility of the formation of supramolecular structures on basis of heterobimetallic, heteroligand triple complexes $Ni^{II}(acac)_2 \cdot LiSt(NaSt) \cdot PhOH$ with the assistance of intermolecular H-bonds.

We have shown that the self-assembly driven growth seems to be because of H-bonding of structures on the basis of $Ni^{II}(acac)_2 \cdot LiSt(NaSt) \cdot PhOH$ with a surface of modified silicone, and further formation supramolecular nanostructures $\{Ni^{II}(acac)_2 \cdot LiSt(NaSt) \cdot PhOH\}_n$ because of directional intermolecular (phenol–carboxylate) H-bonds, and, possibly other noncovalent interactions (van Der Waals-attractions and π-bonding).

These data support the very probable supramolecular structures appearance, on the basis of heterobimetallic, heteroligand triple complexes $Ni^{II}(acac)_2 \cdot LiSt(NaSt) \cdot PhOH$ which occur in the course of the ethylbenzene oxidation with dioxygen, catalyzed by three-component catalytic system $\{Ni^{II}(acac)_2$+LiSt(NaSt)+PhOH} and this can be one of the explanations of the high values of conversion of the ethylbenzene oxidation into α-phenylethyl hydroperoxide at selectivity $\boldsymbol{S}_{PEH}$ preservation at level not below $\boldsymbol{S}_{\mathbf{PEH}}$=90 percent in this process. The higher effectivity of heterobimetallic, heteroligand complexes $Ni^{II}(acac)_2 \cdot NaSt \cdot PhOH$, including metalloligand NaSt, as selective catalysts in comparison with $Ni^{II}(acac)_2 \cdot LiSt \cdot PhOH$, seems to be because of formation of more stable supramolecular structures $\{Ni^{II}(acac)_2 \cdot NaSt \cdot PhOH\}_n$ during hydrocarbon oxidation.

ABBREVIATIONS

AFM method—Atomic-Force Microscopy method
$(Acac)^-$—Acetylacetonate ion
Hacac—acetyl acetone
MP—N-methylpirrolidon-2
MSt—stearates of alkaline metals (M = Li, Na)
RH-Refined hydrocarbon.

KEYWORDS

- **α-phenylethyl hydroperoxide**
- **Bimetallic heteroligand complexes NiII(acac)2**
- **Catalysis**
- **Dioxygen**
- **Ethylbenzene**
- **H-bonds**
- **LiSt(NaSt)·PhOH**
- **Nanostructures**
- **Oxidation**

REFERENCES

1. Suresh, A. K.; Sharma, M. M.; and Sridhar, T.; Industrial Hydrocarbon Oxidation, *Ind. Eng. Chem. Res*. 2000, 39(11), 3958–3969.
2. Weissermel, K.; and Arpe, H.-J.; Industrial Organic Chemistry. 3nd ed., Transl. by Lindley, C. R.; New York: VCH; **1997**.
3. Matienko, L. I.; Solution of the problem of selective oxidation of alkylarenes by molecular oxygen to corresponding hydroperoxides. Catalysis initiated by Ni(II), Co(II), and Fe(III) complexes activated by additives of electron-donor mono- or multidentate extra-ligands, In: Reactions and Properties of Monomers and Polymers. D'Amore, A.; and Zaikov, G.; eds. Chapter 2, New York: Nova Sience Publication Inc.; 2007, 21–41 pp.
4. Matienko, L. I.; Mosolova, L. A.; and Zaikov, G. E.; Selective Catalytic Hydrocarbons Oxidation. New Perspectives. New York, USA: Nova Science Publication Inc.; 2010, 150 p.

5. Leninger, St.; Olenyuk, B.; Stang, P. J.; Self-assembly of discrete cyclic nanostructures mediated by transition metals. *Chem. Rev.* 2000, *100(3),* 853–908.
6. Stang, P. J.; and Olenyuk, B.; Self-assembly, symmetry, and molecular architecture: Coordination as the motif in the rational design of supramolecular metallacyclic polygons and polyhedra, *Acc. Chem. Res.* 1997, *30(12),* 502–518.
7. Drain, C. M.; and Varotto Radivojevic, A. I.; Self-organized porphyrinic materials. *Chem. Rev.* 2009, *109(5),* 1630–1658.
8. Beletskaya, I.; Tyurin, V. S.; Yu., A.; Tsivadze, R.; and Stem, Guilard Ch.; Supramolecular chemistry of metalloporphyrins. *Chem. Rev.* 2009, *109(5),* 1659–1713.
9. Matienko, L. I.; Mosolova, L. A.; Binyukov, V. I.; Mil, E. M.; and Zaikov, G. E.; "The new approach to research of mechanism catalysis with nickel complexes in alkylarens oxidation" "Polymer Yearbook" 2011. New York: Nova Science Publishers; 2012, 221–230 pp.
10. Ludmila Matienko, Vladimir Binyukov, Larisa Mosolova and Gennady Zaikov; The selective ethylbenzene oxidation by dioxygen into α-phenyl ethyl hydroperoxide, catalyzed with triple catalytic system $\{Ni^{II}(acac)_2 + NaSt(LiSt) + PhOH\}$. Formation of nanostructures $\{Ni^{II}(acac)_2 \cdot NaSt \cdot (PhOH)\}_n$ with assistance of intermolecular H-bonds. *Polym. Res. J.* 2011, *5(4),* 423–431.
11. Park, Y. J.; Ziller, J. W.; and Borovik, A. S.; The effect of Redox-Inactive Metal Ions on the Activation of Dioxygen: Isolation and Characterization of Heterobimetallic Complex Containing a Mn^{III}-(μ-OH)-Ca^{II} Core. *J. Am. Chem. Soc.* 2011, *133(24),* 9258–9261.
12. Dubey, M.; Koner, R. R.; and Ray, M.; Sodium and potassium ion directed self-assembled multinuclear assembly of divalent nickel or copper and L-Leucine derived ligand. *Inorg. Chem.* 2009, *48(19),* 9294–9302.
13. Basiuk, E. V.; Basiuk, V. V.; Gomez-Lara, J.; and Toscano, R. A.; A bridged high-spin complex bis-[Ni(II)(rac-5,5,7,12,12,14-hexamethyl-1,4,8,11-tetraazacyclotetradecane)]-2,5-pyridinedicaboxylate diperchlorate monohydrate. *J. Incl. Phenom. Macrocycl. Chem.* 2000, *38(1),* 45–56.
14. Mukherjee, P.; Drew, M. G. B.; Gómez-Garcia, C. J.; and Ghosh, A.; (Ni_2), (Ni_3), and ($Ni_2 + Ni_3$): A unique example of isolated and cocrystallized Ni_2 and Ni_3 Complexes, *Inorg. Chem.* 2009, *48(11),* 4817–4825.

CHAPTER 19

POLYMER-BASED MEMBRANES: FROM INTRODUCTION TO APPLICATION

A. K. HAGHI and G. E. ZAIKOV

CONTENTS

19.1 MEMBRANES FILTRATION

Membrane filtration is a mechanical filtration technique which uses an absolute barrier to the passage of particulate material like any technology currently available in water treatment. The term "membrane" covers a wide range of processes, including those used for gas/gas, gas/liquid, liquid/liquid, gas/solid, and liquid/solid separations. Membrane production is a large-scale operation. There are two basic types of filters: depth filters and membrane filters.

Depth filters have a significant physical depth and the particles to be maintained are captured throughout the depth of the filter. Depth filters often have a flexuous three-dimensional structure, with multiple channels, and heavy branching such that there is a large pathway through which the liquid must flow and by which the filter can retain particles. Depth filters have the advantages of low cost, high throughput, large particle retention capacity, and the ability to retain a variety of particle sizes. However, they bear entrainment of the filter medium, uncertainty regarding effective pore size, some ambiguity regarding the overall integrity of the filter, and the risk of particles being mobilized when the pressure differential across the filter is large.

The second type of filter is the membrane filter, in which depth is not considered momentous. The membrane filter uses a relatively thin material with a well-defined maximum pore size and the particle retaining effect takes place almost entirely at the surface. Membranes offer the advantage of having well-defined effective pore sizes, can be integrity tested more easily than depth filters, and can achieve more filtration of much smaller particles. They tend to be more expensive than depth filters and usually cannot achieve the throughput of a depth filter. Filtration technology has developed a well-defined terminology that has been well addressed by commercial suppliers.

The term membrane has been defined in a number of ways. The most appealing definitions to us are the following:

"A selective separation barrier for one or several components in solution or suspension" [19]. "A thin layer of material that is capable of separating materials as a function of their physical and chemical properties when a driving force is applied across the membrane."

Membranes are important materials which form part of our daily lives. Their long history and use in biological systems has been extensively studied throughout the scientific field. Membranes have proven themselves as promising separation candidates because of advantages offered by their high stability, efficiency, low energy requirement, and ease of operation. Membranes with good thermal and mechanical stability combined with good solvent resistance are important for industrial processes [1].

The concept of membrane processes is relatively simple but nevertheless often unknown. Membranes might be described as conventional filters but with much finer mesh or much smaller pores to enable the separation of tiny particles, even molecules. In general, one can divide membranes into two groups: Porous and nonporous. The former group is similar to classical filtration with pressure as the driving force; the separation of a mixture is achieved by the rejection of at least one component by the membrane and passing of the other components through the membrane (see Figure 19.1). However, it is important to note that nonporous membranes do not operate on a size exclusion mechanism.

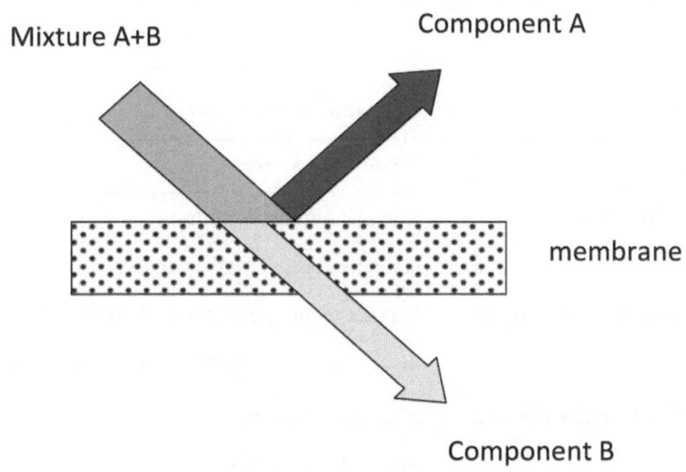

FIGURE 19.1 Basic principle of porous membrane processes.

Membrane separation processes can be used for a wide range of applications and can often offer significant advantages over conventional separation such as distillation and adsorption since the separation is based on a physical mechanism. Compared to conventional processes, therefore, no chemical, biological, or thermal change of the component is involved for most membrane processes. Hence membrane separation is particularly attractive to the processing of food, beverage, and bioproducts where the processed products can be sensitive to temperature (versus distillation), and solvents (vs. extraction).

Synthetic membranes show a large variety in their structural forms. The material used in their production determines their function and their driving forces. Typically the driving force is pressure across the membrane barrier (see Table 19.1) [2–4]]. Formation of a pressure gradient across the membrane allows separation in a bolter-like manner. Some other forms of separation that exist include charge effects and solution diffusion. In this separation, the smaller particles are allowed to pass through as permeates whereas the larger molecules (macromolecules) are retained. The retention or permeation of these species is determined by the pore architecture as well as pore sizes of the membrane employed. Therefore based on the pore sizes, these pressure driven membranes can be divided into reverse osmosis (RO), nanofiltration (NF), ultrafiltration (UF), and microfiltration (MF), which are already applied on an industrial scale to food and bioproduct processing [5–7].

TABLE 19.1 Driving forces and their membrane processes

Driving Force	Membrane Process
Pressure difference	Microfiltration, ultrafiltration
	Nanofiltration, Reverse osmosis
Chemical potential difference	Pervaporation, pertraction, dialysis, gas
	separation, vapor permeation, liquid membranes
Electrical potential difference	Electrodialysis, membrane electrophoresis
	Membrane electrolysis
Temperature difference	Membrane distillation

(i) Microfiltration membranes

Microfiltration (MF) membranes have the largest pore sizes and thus use less pressure. They involve removing chemical and biological species with diameters ranging between 100 and 10,000 nm and components smaller than this, pass through as permeates. MF is primarily used to separate particles and bacteria from other smaller solutes [4].

(ii) Ultrafiltration membranes

Ultrafiltration (UF) membranes operate within the parameters of the micro- and nanofiltration membranes. Therefore UF membranes have smaller pores as compared to MF membranes. They involve retaining macromolecules and colloids from solution which range from 2 to 100 nm and operating pressures between 1 and 10 bar for example, large organic molecules and proteins. UF is used to separate colloids such as proteins from small molecules such as sugars and salts [4].

(iii) Nanofiltration membranes

Nanofiltration (NF) membranes are distinguished by their pore sizes of from 0.5 to 2 nm and operating pressures between 5 and 40 bar. They are mainly used for the removal of small organic molecules and di- and multivalent ions. Additionally, NF membranes have surface charges that make them suitable for retaining ionic pollutants from solution. NF is used to achieve separation between sugars, other organic molecules, and multivalent salts on the one hand, from monovalent salts and water on the other. Nanofiltration, however, does not remove dissolved compounds [4].

(iv) Reverse osmosis membranes

Reverse osmosis *(RO) membranes are dense semipermeable membranes mainly used for desalination of sea water [38]. Contrary to MF and UF membranes, RO membranes have no distinct pores. As a result, high pressures are applied to increase the permeability of the membranes [4]. The properties of the various types of membranes are summarized in Table 19.2.*

TABLE 19.2 Summary of properties of pressure driven membranes [4]

	MF	UF	NF	RO
Permeability(L/h.m^2.bar)	1,000	10–1,000	1.5–30	0.05–1.5
Pressure (bar)	0.1–2	0.1–5	3–20	5–1,120
Pore size (nm)	100–10,000	2–100	0.5–2	0.5
Separation Mechanism	sieving	sieving	Sieving, charge effects	Solution diffusion
Applications	Removal of bacteria	Removal of bacteria, fungi, virses	Removal of multivalentions	Desalinatiob

The NF membrane is a type of pressure-driven membrane with properties in between RO and UF membranes. NF offers several advantages such as low operation pressure, high flux, high retention of multivalent anion salts and an organic molecular above 300, relatively low investment, and low operation and maintenance costs. Because of these advantages, the applications of NF worldwide have increased [8]. In recent times, research in the application of nanofiltration techniques has been extended from separation of aqueous solutions to separation of organic solvents to homogeneous catalysis, separation of ionic liquids, food processing, and so on [9].

Figure 19.2 presents a classification on the applicability of different membrane separation processes based on particle or molecular sizes. RO process is often used for desalination and pure water production, but it is the UF and MF that are widely used in food and bioprocessing.

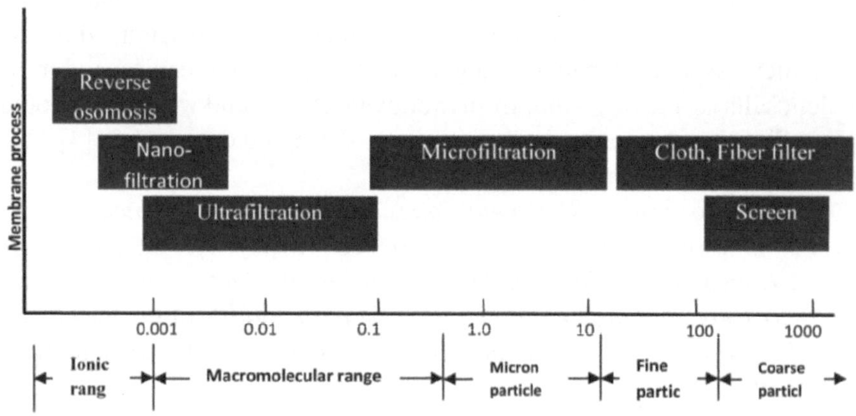

FIGURE 19.2 The applicability ranges of different separation processes based on sizes.

As MF membranes target on the microorganism removal, they are given the absolute rating, namely for the diameter of the largest pore on the membrane surface. UF/NF membranes are characterized by the nominal rating because of their early applications of purifying biological solutions. The nominal rating is defined as the molecular weight cut-off (MWCO) that is the smallest molecular weight of species, of which the membrane has more than 90 percent rejection (see later for definitions). The separation mechanism in MF/UF/NF is mainly the size exclusion, which is indicated in the nominal ratings of the membranes. The other separation mechanism includes the electrostatic interactions between solutes and membranes, which depends on the surface and physiochemical properties of solutes and membranes [5]. Also, the principal types of membrane are shown schematically in Figure 19.4 and 19.3 and are described briefly below.

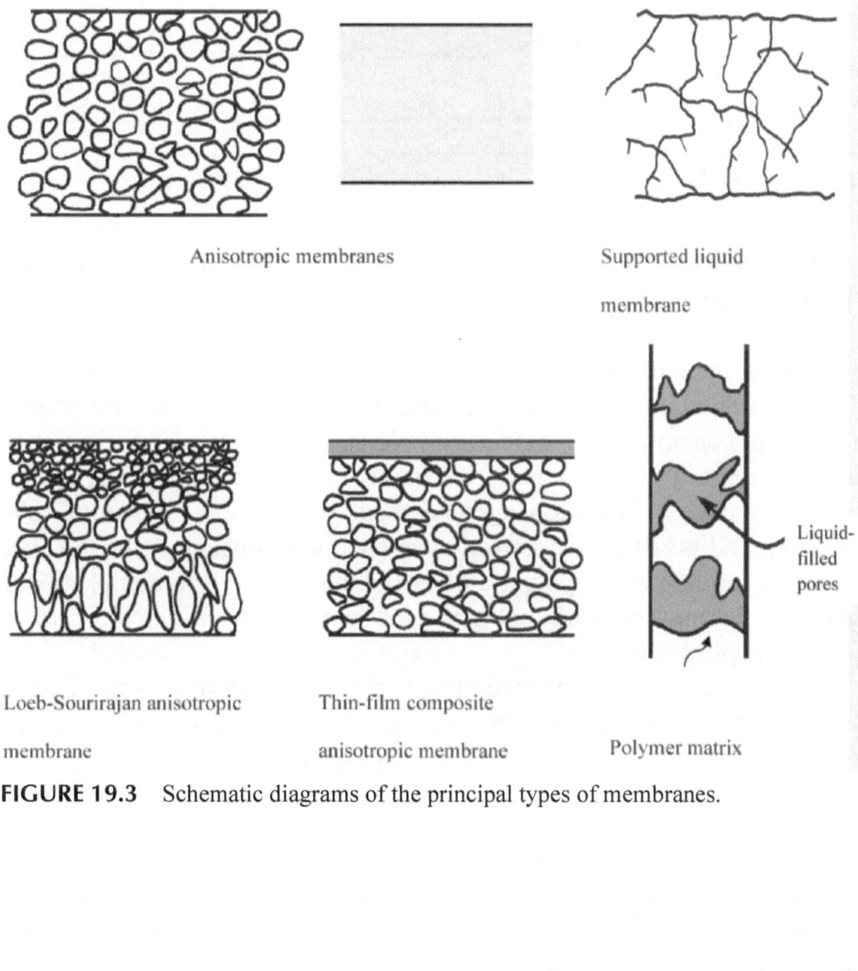

FIGURE 19.3 Schematic diagrams of the principal types of membranes.

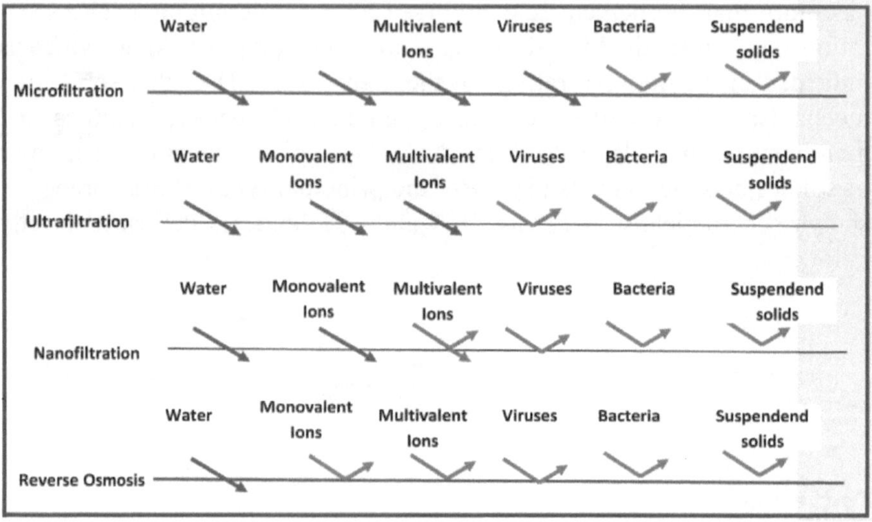

FIGURE 19.4 Membrane process characteristics.

19.2 THE RELATIONSHIP BETWEEN NANOTECHNOLOGY AND FILTRATION

Nowadays, nanomaterials have become the most interesting topic of materials research and development because of their unique structural properties (unique chemical, biological, and physical properties as compared to larger particles of the same material) that cover their efficient uses in various fields, such as ion exchange and separation, catalysis, biomolecular isolation, and purification, as well as in chemical sensing [10]. However, the understanding of the potential risks (health and environmental effects) posed by nanomaterials has not increased as rapidly as research regarding possible applications.

One of the ways to enhance their functional properties is to increase their specific surface area by the creation of a large number of nanostructured elements or by the synthesis of a highly porous material.

Classically, porous matter is seen as material containing three-dimensional voids, representing translational repetition, while no regularity is necessary for a material to be termed "porous." In general, the pores can be classified into two types: Open pores which connect to the surface of the material, and closed pores which are isolated from the outside. If the material exhibits mainly open pores, which can be easily transpired, then one can consider its use in functional applications such as adsorption, catalysis, and sensing. In turn, the closed pores can be used in sonic and thermal insulation, or lightweight structural applications. The use of porous materials offers also new opportunities, including areas in chemistry such as the guest–host synthesis, the molecular manipulations, and in the reactions for manufacture of nanoparticles, nanowires, and other quantum nanostructures. The International Union of Pure and Applied Chemistry (IUPAC) defines porosity scales as follows (Figure 19.5):

- Microporous materials 0–2 nm pores
- Mesoporous materials 2–50 nm pores
- Macroporous materials >50 nm pores

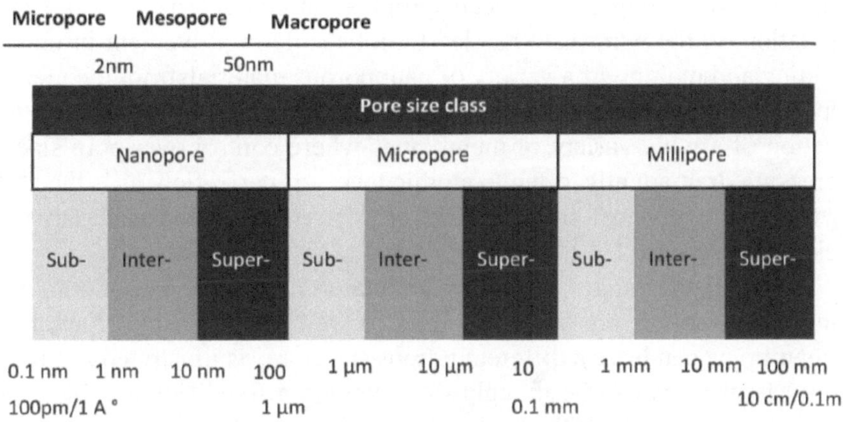

FIGURE 19.5 New pore size classification as compared with the current IUPAC nomenclature.

This definition, it should be noted, is somewhat in conflict with the definition of nanoscale objects, which typically have large relative porosities (>0.4), and pore diameters between 1 and 100 nm. In order to classify porous materials according to the size of their pores the sorption analysis is one of the tools often used. This tool is based on the fact that pores of different sizes lead to totally different characteristics in sorption isotherms. The correlation between the vapor pressure and the pore size can be written as per the Kelvin equation:

$$r_p\left(\frac{p}{p_0}\right)=\frac{2\gamma V_L}{RT\ln\left(p/p_0\right)}+t\left(\frac{p}{p_0}\right) \tag{19.1}$$

Therefore, the isotherms of microporous materials show a steep increase at very low pressures (relative pressures near zero) and reach aplateau quickly. Mesoporous materials are characterized by a so called capillary doping step and hysteresis (a discrepancy between adsorption and desorption). Macroporous materials show a single or multiple adsorption steps near the pressure of the standard bulk condensed state (relative pressure approaches one) [10].

Nanoporous materials exuberate in nature, both in biological systems and in natural minerals. Some nanoporous materials have been used in industries for a long time. Recent progress in characterization and manipulation on the nanoscale has led to noticeable improvement in understanding and making of a variety of nanoporous materialsfrom the merely opportunistic to directed design. This is most strikingly the case in the creation of a wide variety of membranes where control over pore size is increasing dramatically, often to atomic levels of perfection, as is the ability to modify physical and chemical characteristics of the materials that make up the pores [11].

The available ranges of membrane materials include polymeric, carbon, silica, zeolite, and other ceramics, as well as composites. Each type of membrane can have a different porous structure, as illustrated in Figure 19.6. Membranes can be thought of as having a fixed (immovable) network of pores in which the molecule travels, with the exception of most polymeric membranes [12–13]. Polymeric membranes are composed of an amorphous mix of polymer chains whose interactions involve mostly van der Waals forces. However, some polymers manifest a behavior that is consistent with the idea of existence of open pores within their matrix. This is especially true for high free volume, high permeability polymers, as has been proved by computer modeling, low activation energy of diffusion, negative activation energy of permeation, solubility controlled permeation [14–15]. Although polymeric membranes have often been viewed as nonporous, in the modeling framework discussed here, it is convenient to consider them nonetheless as porous. Glassy polymers have pores that can be considered as "frozen" over short time scales, while rubbery polymers have dynamic fluctuating pores (or more correctly free volume elements) that move, shrink, expand, and disappear [16].

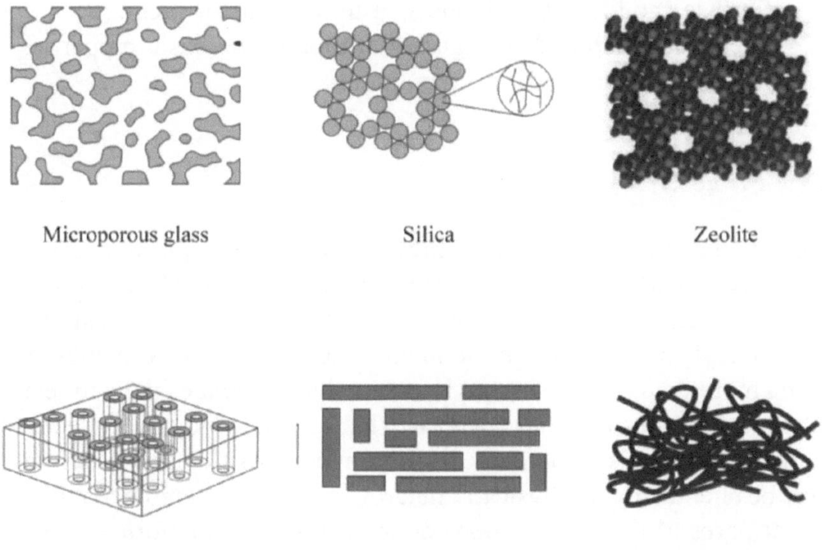

FIGURE 19.6 Porous structure within various types of membranes.

Three nanotechnologies, often used in the filtering processes, and showing great potential for applications in remediation are:

1. Nanofiltration (and its sibling technologies: reverse osmosis, ultrafiltration, and microfiltration), is a fully-developed, commercially-available membrane technology with a large number of vendors. Nanofiltration relies on the ability of membranes to discriminate between the physical size of particles or species in a mixture or solution and is primarily used for water pretreatment, treatment, and purification). There are almost 600 companies worldwide offering membrane systems.
2. Electrospinning is a process utilized by the nanofiltration process, in which fibers are stretched and elongated down to a diameter of about 10 nm. The modified nanofibers that are produced are particularly useful in the filtration process as an ultra-concentrated filter with a very large surface area. Studies have found that electrospun nanofibers can capture metallic ions and are continually effective through refiltration.

3. Surface modified membrane is a term used for membranes with altered makeup and configuration, though the basic properties of their underlying materials remain intact.

19.3 TYPES OF MEMBRANES

As mentioned, membranes have achieved a momentous place in chemical technology and are used in a broad range of applications. The key property that is exploited is the ability of a membrane to control the permeation rate of a chemical species through the membrane. In essence, a membrane is nothing more than a discrete, thin interface that moderates the permeation of chemical species in contact with it. This interface may be molecularly homogeneous, that is completely uniform in composition and structure, or it may be chemically or physically heterogeneous for example, containing holes or pores of finite dimensions or consisting of some form of layered structure. A normal filter meets this definition of a membrane, but, generally, the term filter is usually limited to structures that separate particulate suspensions larger than 1–10 μm [17].

The preparation of synthetic membranes is, however, a more recent invention which has received a great audience because of its applications [18]. Membrane technology like most other methods has undergone a developmental stage, which has validated the technique as a cost-effective treatment option for water. The level of performance of the membrane technologies is still developing and it is stimulated by the use of additives to improve the mechanical and thermal properties, as well as the permeability, selectivity, rejection, and fouling of the membranes [19]. Membranes can be fabricated to possess different morphologies. However, most membranes that have found practical use are mainly of asymmetric structure. Separation in membrane processes takes place as a result of differences in the transport rates of different species through the membrane structure, which is usually polymeric or ceramic [20].

The versatility of membrane filtration has allowed their use in many processes where their properties are suitable in the feed stream. Although membrane separation does not provide the ultimate solution to water treatment, it can be economically connected to conventional treatment technologies by modifying and improving certain properties [21].

The performance of any polymeric membrane in a given process is highly dependent on both the chemical structure of the matrix and the physical arrangement of the membrane [22]. Moreover, the structural integrity of a membrane is very important since it determines its permeation and selectivity efficiency. As such, polymer membranes should be seen as much more than just sieving filters, but as intrinsic complex structures which can either be homogenous (isotropic) or heterogeneous (anisotropic), porous or dense, liquid or solid, organic or inorganic [22–23].

19.3.1 ISOTROPIC MEMBRANES

Isotropic membranes are typically homogeneous/uniform in composition and structure. They are divided into three subgroups, namely: Microporous, dense, and electrically charged membranes [20]. Isotropic microporous membranes have evenly distributed pores (Figure 19.7a) [27]. Their pore diameters range between 0.0 and 10 μm and operate by the sieving mechanism. The microporous membranes are mainly prepared by the phase inversion method albeit other methods can be used. Conversely, isotropic dense membranes do not have pores and as a result they tend to be thicker than the microporous membranes (Figure 19.7b). Solutes are carried through the membrane by diffusion under pressure, concentration or electrical potential gradient. Electrically charged membranes can either be porous or nonporous. However in most cases they are finely microporous with pore walls containing charged ions (Figure 19.7c) [20, 28].

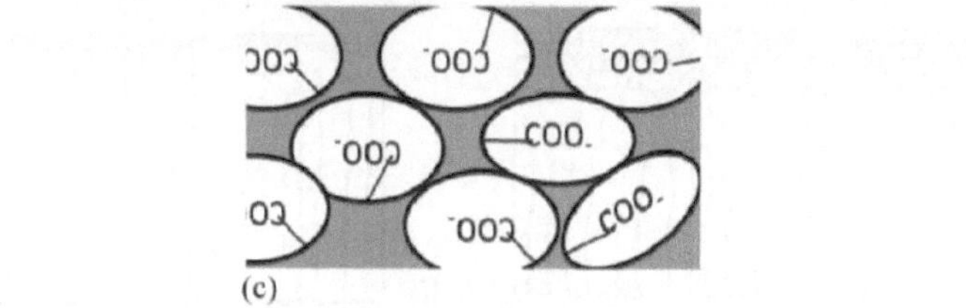

FIGURE 19.7 Schematic diagrams of isotropic membranes: (a) Microporous; (b) dense; and (c) electrically charged membranes.

19.3.2 ANISOTROPIC MEMBRANES

Anisotropic membranes are often referred to as Loeb-Sourirajan, based on the scientists who first synthesized them [24–25]. They are the most widely used membranes in industries. The transport rate of a species through a membrane is inversely proportional to the membrane thickness. The membrane should be as thin as possible, and with high transport rates to be eligible in membrane separation processes for economic reasons. Contractual film fabrication technology limits manufacture of mechanically strong, defect-free films to thicknesses of about 20 μm. The development of novel membrane fabrication techniques to produce anisotropic membrane structures is one of the major breakthroughs of membrane technology. Anisotropic membranes consist of an extremely thin surface layer supported on a much thicker, porous substructure. The surface layer and its substructure may be formed in a single operation or separately [17]. They are represented by nonuniform structures which consist of a thin active skin layer and a highly porous support layer. The active layer enjoins the efficiency of the membrane, whereas the porous support layer influences the mechanical stability of the membrane. Anisotropic membranes can be classified into two groups, namely: (1) integrally skinned membranes where the active layer is formed from the same substance as the supporting layer and (2) composite membranes where the polymer of the active layer differs from that of the supporting sublayer [25]. In composite membranes, the layers are usually made from different polymers. The separation properties and permeation rates of the membrane are determined particularly by the surface layer, and the substructure functions as a mechanical support. The advantages of the higher fluxes provided by anisotropic membranes are so great that almost all commercial processes use such membranes [17] (Figure 19.8).

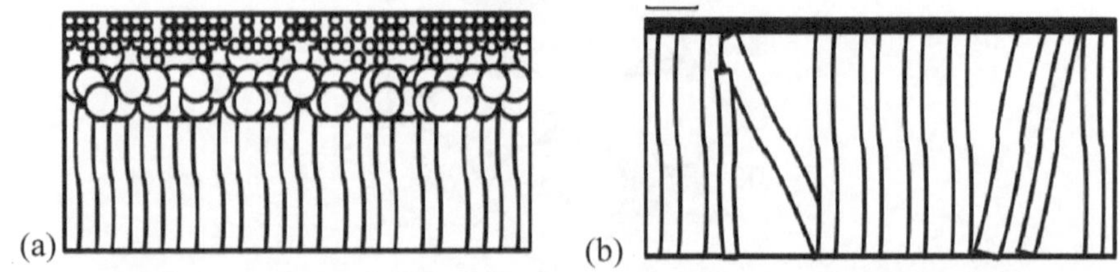

FIGURE 19.8 Schematic diagrams of anisotropic membranes: (a) Loeb-Sourirajan and (b) thin film composite membranes.

19.3.3 POROUS MEMBRANE

In Knudsen diffusion (Figure 19.9a), the pore size forces the penetrant molecules to collide more frequently with the pore wall than with other incisive species [26]. Except for some special applications as membrane reactors, Knudsen-selective membranes are not commercially attractive because of their low selectivity [27]. In surface diffusion mechanism (Figure 19.9b), the pervasive molecules adsorb on the surface of the pores and so move from one site to another of lower concentration. Capillary condensation (Figure 19.9c) impresses the rate of diffusion across the membrane. It occurs when the pore size and the interactions of the penetrant with the pore walls induce penetrant condensation in the pore [28]. Molecular-sieve membranes in Figure 19.9(d) have received more attention because of their higher productivities and selectivity than solution-diffusion membranes. Molecular sieving membranes are means to polymeric membranes. They have ultra microporous (<7 Å) with sufficiently small pores to barricade some molecules, while allowing others to pass through. Although they have several advantages such as permeation performance, chemical and thermal stability, they are still difficult to process because of some properties like fragility. Also they are expensive to fabricate.

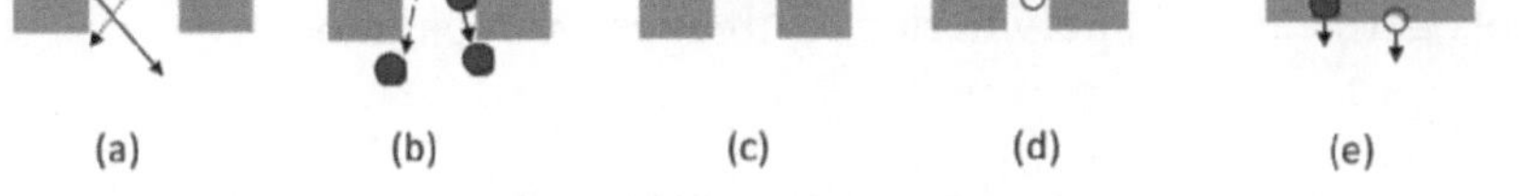

FIGURE 19.9 Schematic representation of membrane-based gas separations: (a) Knudsen-flow separation, (b) surface-diffusion, (c) capillary condensation, (d) molecular-sieving separation, and (e) solution-diffusion mechanism.

19.3.4 NONPOROUS (DENSE) MEMBRANE

Nonporous, dense membranes consist of a dense film through which permeates are transported by diffusion under the driving force of a pressure, concentration, or electrical potential gradient. The separation of various components of a mixture is related directly to their relative transport rate within the membrane, which is determined by their diffusivity and solubility in the membrane material. Thus, nonporous, dense membranes can separate permeates of similar size if the permeate concentrations in the membrane material differ substantially. Reverse osmosis membranes use dense membranes to perform the separation. Usually these membranes have an anisotropic structure to improve the flux [17].

The mechanism of separation by nonporous membranes is different from that by porous membranes. The transport through nonporous polymeric membranes is usually described by a solution–diffusion mechanism (Figure 19.9e). The most current commercial polymeric membranes operate according to the solution–diffusion mechanism. The solution–diffusion mechanism has three steps: (1) absorption or adsorption at the upstream boundary, (2) activated diffusion through the membrane, and (3) desorption or evaporation on the other side. This solution–diffusion mechanism is driven by a difference in the thermodynamic activities existing at the upstream and downstream faces of the membrane as well as the intermolecular forces acting between the permeating molecules and those making up the membrane material.

The concentration gradient causes diffusion in the direction of decreasing activity. Differences in the permeability in dense membranes are caused not only by diffusivity differences of the various species, but also by differences in the physicochemical interactions of the species within the polymer. The solution–diffusion model assumes that the pressure within a membrane is uniform and that the chemical potential gradient across the membrane is expressed only as a concentration gradient. This mechanism controls permeation in polymeric membranes for separations.

19.4 CARBON NANOTUBES-POLYMER MEMBRANE

Iijima discovered carbon nanotubes (CNTs) in 1991 and it was really a revolution in nanoscience because of their distinguished properties. Carbon nanotubes have the unique electrical properties and extremely high thermal conductivity [29–30] and high elastic modulus (>1 TPa), with large elastic strain-up to 5 percent, and large breaking strain-up to 20 percent. Their excellent mechanical properties could lead to many applications [31]. For example, with their amazing strength and stiffness, and with the advantage of lightness, prospective future applications of CNTs are in aerospace engineering and virtual biodevices [32].

Carbon nanotubes have been studied worldwide by scientists and engineers since their discovery, but a robust, theoretically precise and efficient prediction of the mechanical properties of CNTs has not yet been found. The problem is, when the size of an object is nanoscale in size, their many physical properties cannot be modeled and analyzed by using constitutive laws from traditional continuum theories, since the complex atomistic processes affect the results of their macroscopic behavior. Atomistic simulations can give more precise modeled results of the underlying physical properties. Because atomistic simulations of a whole CNT are computationally infeasible at present, a new atomistic and continuum mixing modeling method is needed to solve the problem, which requires crossing the length and time scales. The research here is to develop a proper technique of spanning multiscales from atomic to macroscopic space, in which the constitutive laws are derived from empirical atomistic potentials which deal with individual interactions between single atoms at the micro-level, whereas Cosserat continuum theories are adopted for a shell model through the application of the Cauchy-Born rule to give the properties which represent the averaged behavior of large volumes of atoms at the macro-level [33–34]. Since experiments of CNTs are relatively expensive at present, and often unexpected manual errors could be involved, it will be very helpful to have a mature theoretical method for the study of mechanical properties of CNTs. Thus, if this research is successful, it could also be a reference for the research of all sorts of research at the nanoscale, and the results can be of interest to aerospace, biomedical engineering [35].

Subsequent investigations have shown that CNTs integrate amazing rigid and tough properties, such as exceptionally high elastic properties, large elastic strain, and fracture strain sustaining capability, which seem inconsistent and impossible in the previous materials. Carbon nanotubes are the strongest fibers known. The Young's Modulus of SWNT is around 1 TPa, which is 5 times greater than steel (200 GPa) while the density is only 1.2–1.4 g/cm^3. This means that materials made of nanotubes are lighter and more durable.

Beside their well-known extra-high mechanical properties, single-walled carbon nanotubes (SWNTs) offer either metallic or semiconductor characteristics based on the chiral structure of fullerene. They possess superior thermal and electrical properties so SWNTs are regarded as the most promising reinforcement material for the next generation of high performance structural and multifunctional composites, and evoke great interest in polymer-based composites research. The SWNTs/polymer composites are theoretically predicted to have both exceptional mechanical and functional properties, which carbon fibers cannot offer [36].

19.4.1 CARBON NANOTUBES

Nanotubular materials are important "building blocks" of nanotechnology, in particular, the synthesis and applications of CNTs [37–39]. One application area has been the use of carbon nanotubes for molecular separations, owing to some of their unique properties. One such important property is extremely fast mass transport of molecules within carbon nanotubes associated with their low friction inner nanotube surfaces, which has been demonstrated via computational and experimental studies [40–41]. Furthermore, the behavior of adsorbate molecules in nano-confinement is fundamentally different than in the bulk phase, which could lead to the design of new sorbents [42].

Finally, their one-dimensional geometry could allow for alignment in desirable orientations for given separation devices to optimize the mass transport. Despite possessing such attractive properties, several intrinsic limitations of carbon nanotubes inhibit their application in large scale separation processes such as the high cost of CNT synthesis and membrane formation (by microfabrication processes), as well as their lack of surface functionality, which significantly limits their molecular selectivity [43]. Although outer-surface modification of carbon nanotubes has been developed for nearly two decades, interior modification via covalent chemistry is still challenging because ofthe low reactivity of the inner-surface. Specifically, forming covalent bonds at inner walls of carbon nanotubes requires a transformation from sp^2 to sp^3 hybridization. The formation of sp^3 carbon is energetically unfavorable for concave surfaces [44].

Membrane is a potentially effective way to apply nanotubular materials in industrial-scale molecular transport and separation processes. Polymeric membranes are already prominent for separations applications because of their low fabrication and operation costs. However, the main challenge for utilizing polymer membranes for future high-performance separations is to overcome the tradeoff between permeability and selectivity. A combination of the potentially high throughput and selectivity of nanotube materials with the process ability and mechanical strength of polymers may allow for the fabrication of scalable, high-performance membranes [45–46].

19.4.2 STRUCTURE OF CARBON NANOTUBES

Two types of nanotubes exist in nature–one is multi-walled carbon nanotube (MWNTs(, which were discovered by Iijima in 1991 [39] and the other SWNTs, which were discovered by Bethune et al. in 1993 [47–48].

Single-wall nanotube has only one single layer with diameters in the range of 0.6–1 nm and densities of 1.33–1.40 g/cm^3 [49] MWNTs are simply composed of concentric SWNTs with an inner diameter from 1.5 to 15 nm and the outer diameter from 2.5 nm to 30 nm [50]. SWNTs have better defined shapes of cylinder than MWNT, thus MWNTs have more possibilities of structure defects and their nanostructure being less stable. Their specific mechanical and electronic properties make them useful for future high strength/modulus materials and nanodevices. They exhibit low density, large elastic limit without breaking (up to 20–30% strain before failure), exceptional elastic stiffness, greater than 1,000 GPa, and their extreme strength which is more than 20 times higher than a high-strength steel alloy. Besides, they also posses superior thermal and elastic properties with thermal stability up to 2,800°C in vacuum and up to 750°C in air, thermal conductivity about twice as high as diamond, electric current carrying capacity 1,000 times higher than copper wire [51]. The properties of CNTs strongly depend on the size and the chirality and dramatically change when SWCNTs or MWCNTs are considered [52].

CNTs are formed from pure carbon bonds. Pure carbons only have two covalent bonds: sp^2 and sp^3. The former constitutes graphite and the latter constitutes diamond. The sp^2 hybridization, composed of one's orbital and two p orbitals, is a strong bond within a plane but weak between planes. When more bonds come together, they form six-fold structures, like honeycomb pattern, which is a plane structure, similar as graphite [53].

Graphite is stacked layer by layer so it is only stable for one single sheet. Wrapping these layers into cylinders and joining the edges, a tube of graphite is formed, called nanotube [54].

Atomic structure of nanotubes can be described in terms of tube chirality, or helicity, which is defined by the chiral vector, and the chiral angle, θ. Figure 19.10 shows visuallay cutting a graphite sheet along the dotted lines and rolling the tube so that the tip of the chiral vector touches its tail. The chiral vector, often known as the roll-up vector, can be described by the following equation [55]:

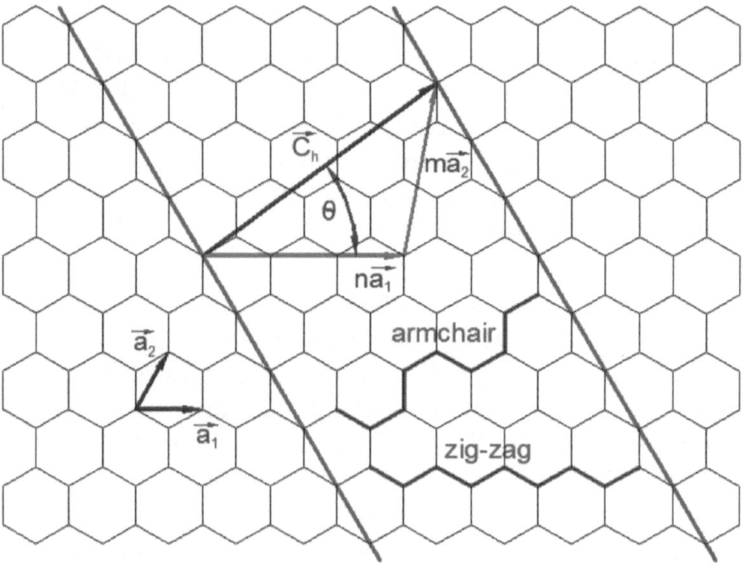

FIGURE 19.10 Schematic diagram showing how graphite sheet is "rolled" to form CNT.

$$C_h = na_1 + ma_2 \tag{19.2}$$

As shown in Figure 19.10, the integers (n, m) are the number of steps along the carbon bonds of the hexagonal lattice. Chiral angle determines the amount of "twist" in the tube. Two limiting cases exist where the chiral angle is at 0° and 30°. These limiting cases are referred to as ziz-zag (0°) and armchair (30°), based on the geometry of the carbon bonds around the circumference of the nanotube. The difference in armchair and zig-zag nanotube structures is shown in Figure 19.11. In terms of the roll-up vector, the ziz-zag nanotube is (n, 0) and the armchair nanotube is (n, n). The roll-up vector of the nanotube also defines the nanotube diameter since the inter-atomic spacing of the carbon atoms is known [36].

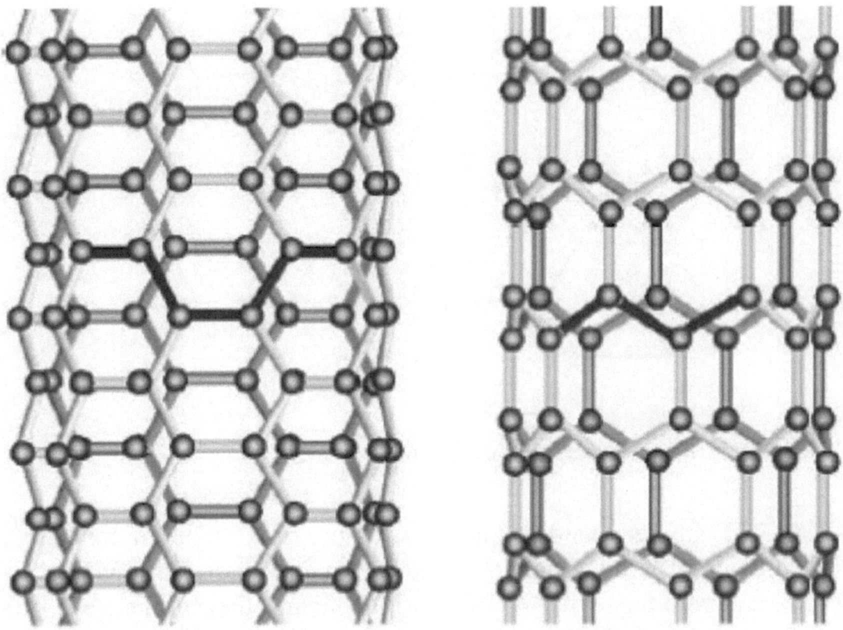

(a) (b)

FIGURE 19.11 Illustrations of the atomic structure (a) an armchair and (b) a ziz-zag nanotube.

Chiral vector C_h is a vector that maps an atom of one end of the tube to the other. C_h can be an integer multiple a_1 of a_2, which are two basis vectors of the graphite celland we have $C_h = a_1 + a_2$, with integer n and m, and the constructed CNT is called a (n, m) CNT, as shown in Figure 19.12. It can be proved that for armchair CNTs n = m, and for zigzag CNTs m=0. In Figure 19.12, the structure is designed to be a (4, 0) zigzag SWCNT.

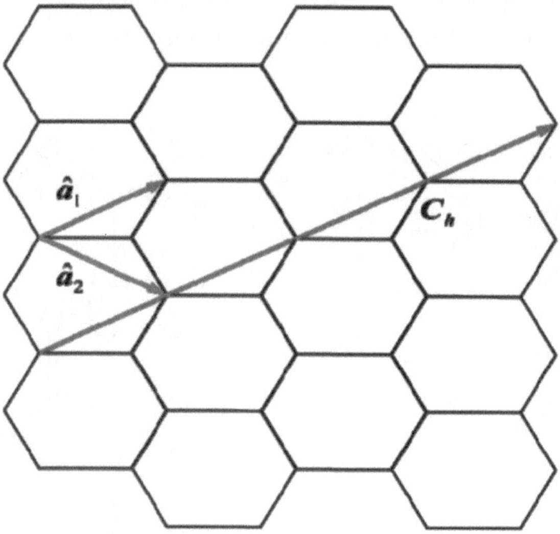

FIGURE 19.12 Basis vectors and chiral vector.

MWCNT can be considered as the structure of a bundle of concentric SWCNTs with different diameters. The length and diameter of MWCNTs are different from those of SWCNTs, which means, their properties differ significantly. MWCNTs can be modeled as a collection of SWCNTs, provided the interlayer interactions are modeled by van der Waals forces in the simulation. A SWCNT can be modeled as a hollow cylinder by rolling a graphite sheet as presented in Figure 19.13.

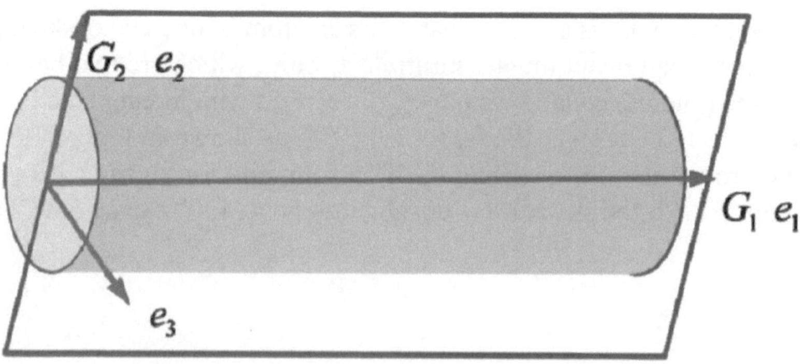

FIGURE 19.13 Illustration of a graphite sheet rolling to SWCNT.

If a planar graphite sheet is considered to be an undeformed configuration, and the SWCNT is defined as the current configuration, then the relationship between the SWCNT and the graphite sheet can be shown to be:

$$e_1 = G_1, e_2 = \mathrm{R}\sin\frac{\mathrm{G}_2}{\mathrm{R}}, e_3 = \mathrm{R}\cos\frac{\mathrm{G}_2}{\mathrm{R}} - \mathrm{R} \tag{19.3}$$

The relationship between the integer's n, m and the radius of SWCNT is given by:

$$R = a\sqrt{m^2 + mn + n^2}\ /\ 2\pi \tag{19.4}$$

where , and a_0 is the length of a nonstretched C–C bond which is 0.142 nm [56].

As a graphite sheet can be "rolled" into a SWCNT, we can "unroll" the SWCNT to a plane graphite sheet. Since a SWCNT can be considered as a rectangular strip of hexagonal graphite monolayer rolling up to a cylindrical tube, the general idea is that it can be modeled as a cylindrical shell, a cylinder surface, or it can pull-back to be modeled as a plane sheet deforming into curved surface in three-dimensional space. A MWCNT can be modeled as a combination of a series of concentric SWCNTs with interlayer intera-atomic reactions. Provided the continuum shell theory captures the deformation at the macro-level, the inner micro-structure can be described by finding the appropriate form of the potential function which is related to the position of the atoms at the atomistic level. Therefore, the SWCNT can be considered as a generalized continuum with microstructure [35].

19.4.3 CNT COMPOSITES

CNT composite materials has led to significant development in nanoscience and nanotechnology. Their remarkable properties offer the potential for fabricating composites with substantially enhanced physical properties including conductivity, strength, elasticity, and toughness. Effective utilization of CNT in composite applications is dependent on the homogeneous distribution of CNTs throughout the matrix. Polymer-based nanocomposites are being developed for electronics applications such as thin-film capacitors in integrated circuits and solid polymer electrolytes for batteries. Research is being conducted throughout the world targeting the application of carbon nanotubes as materials for use in transistors, fuel cells, big TV screens, ultra-sensitive sensors, high-resolution Atomic Force Microscopy (AFM) probes, super-capacitor, transparent conducting film, drug carrier, catalysts, and composite material. Nowadays, there are more reports on the fluid transport through porous CNTs/polymer membrane.

19.4.4 STRUCTURAL DEVELOPMENT IN POLYMER/CNT FIBERS

The inherent properties of CNT assume that the structure is well preserved (large-aspect-ratio and without defects). The first step toward effective reinforcement of polymers using nano-fillers is to achieve a uniform dispersion of the fillers within the hosting matrix, and this is also related to the as-synthesized nano-carbon structure. Secondly, effective interfacial interaction and stress transfer between CNT and polymer is essential for improved mechanical properties of the fiber composite. Finally, similar to polymer molecules, the excellent intrinsic mechanical properties of CNT can be fully exploited only if an ideal uniaxial orientation is achieved. Therefore, during the fabrication of polymer/CNT fibers, four key areas need to be addressed and understood in order to successfully control the micro-structural development in these composites. These are: (i) CNT pristine structure, (ii) CNT dispersion, (iii) polymer–CNT interfacial interaction, and (iv) orientation of the filler and matrix molecules (Figure 19.14).

Figure 19.14 Four major factors affecting the micro-structural development in polymer/CNT composite fiber during processing [57].

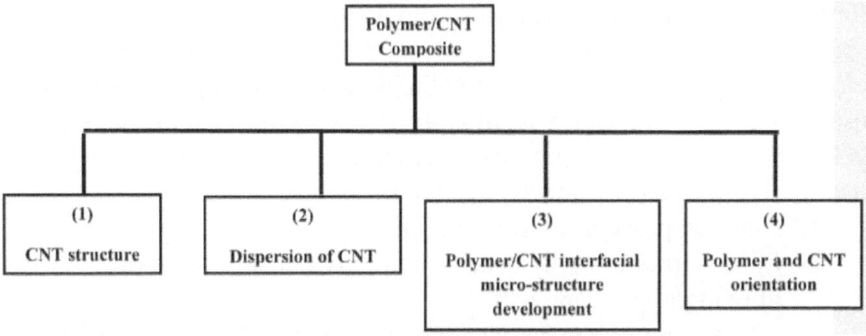

FIGURE 19.14 Four major factors affecting the micro-structural development in polymer/CNT composite fiber during processing.

Achieving homogenous dispersion of CNTs in the polymer matrix through strong interfacial interactions is crucial to the successful development of CNT/polymer nanocomposite [58]. As a result, various chemical or physical modifications can be applied to CNTs to improve its dispersion and compatibility with polymer matrix. Among these approaches acid treatment is considered most convenient, in which hydroxyl and carboxyl groups generated would concentrate on the ends of the CNT and at defect sites, making them more reactive and thus better dispersed [59–60].

The incorporation of functionalized CNTs into composite membranes are mostly carried out on flat sheet membranes [61–62]. For considering the potential influences of CNTs on the physicochemical properties of dope solution [63] and change of membrane formation route originated from various additives [64], it is necessary to study the effects of CNTs on the morphology and performance.

19.4.5 GENERAL FABRICATION PROCEDURES FOR POLYMER/CNT FIBERS

In general, when discussing polymer/CNT composites, two major classes come to mind. First, the CNT nano-fillers, which are dispersed within a polymer at a specified concentration, and this entire mixture fabricated into a composite. Secondly, the grown CNT processed into fibers or films, and this macroscopic CNT materials are then embedded into a polymer matrix [65]. The four major fiber-spinning methods (Figure 19.15) used for polymer/CNT composites from both the solution and melt include dry-spinning [66], wet-spinning [67], dry-jet wet spinning (gel-spinning), and electrospinning [68]. An ancient solid-state spinning approach has been used for fabricating 100 percent CNT fibers from both forests and aero gels. Irrespective of the processing technique, in order to develop high-quality fibers many parameters need to be well controlled.

All spinning procedures generally involve:

(i) Fiber formation, (ii) coagulation/gelation/solidification, and (iii) drawing/alignment.

For all of these processes, the even dispersion of the CNT within the polymer solution or melt is very important. However, in terms of achieving excellent axial mechanical properties, alignment and orientation of the polymer chains and the CNT in the composite is necessary. Fiber alignment is accomplished in postprocessing such as drawing/annealing and is key to increasing crystallinity, tensile strength, and stiffness [69].

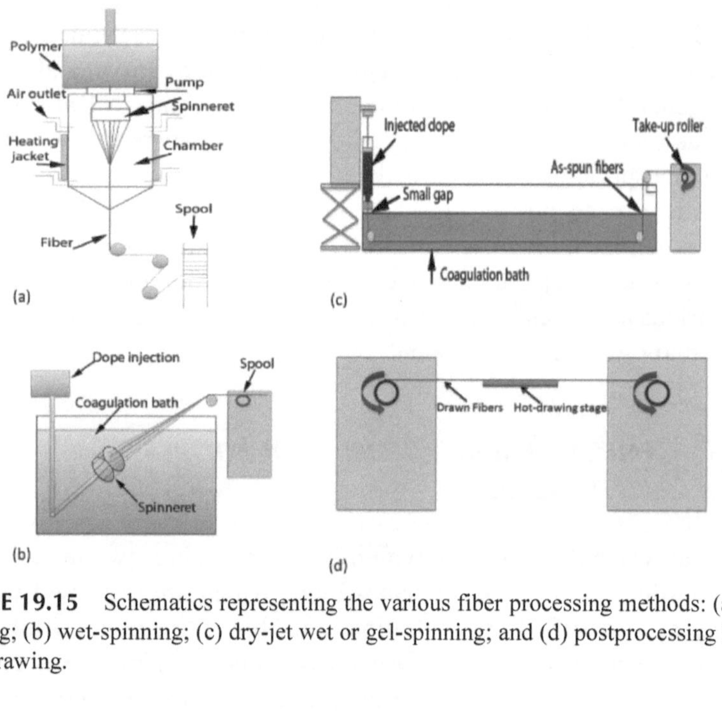

FIGURE 19.15 Schematics representing the various fiber processing methods: (a) Dry-spinning; (b) wet-spinning; (c) dry-jet wet or gel-spinning; and (d) postprocessing by hot-stage drawing.

19.5 COMPUTATIONAL METHODS

Computational approaches to obtain solubility and diffusion coefficients of small molecules in polymers have focused primarily upon equilibrium molecular dynamics (MD) and Monte Carlo (MC) methods. These have been thoroughly reviewed by several investigators [70–71].

Computational approach can play an important role in the development of the CNT-based composites by providing simulation results to help on the understanding, analysis, and design of such nanocomposites. At the nanoscale, analytical models are difficult to establish or are too complicated to solve, and tests are extremely difficult and expensive to conduct. Modeling and simulations of nanocomposites, on the other hand, can be achieved readily and cost effectively on even a desktop computer. Characterizing the mechanical properties of CNT-based composites is just one of the many important and urgent tasks that simulations can follow out [72].

Computer simulations on model systems have in recent years provided much valuable information on the thermodynamic, structural, and transport properties of classical dense fluids. The success of these methods rests primarily on the fact that a model containing a relatively small number of particles is in general found to be sufficient to simulate the behavior of a macroscopic system. Two distinct techniques of computer simulation have been developed which are known as the method of molecular dynamics and the Monte Carlo method [73–75].

Instead of adopting a trial-and-error approach to membrane development, it is far more efficient to have a real understanding of the separation phenomena to guide membrane design [76–79]. Similarly, methods such as MC, MD, and other computational techniques have improved the understanding of the relationships between membrane characteristics and separation properties. In addition to these inputs, it is also beneficial to have simple models and theories that give an overall insight into separation performance [80–83].

19.5.1 PERMEANCE AND SELECTIVITY OF SEPARATION MEMBRANES

A membrane separates one component from another on the basis of size, shape, or chemical affinity. Two characteristics dictate membrane performance, permeability, that is the flux of the membrane, and selectivity or the membrane's preference to pass one species and not another [84].

A membrane can be defined as a selective barrier between two phases, the "selective" being inherent to a membrane or a membrane processes. The membrane separation technology is proving to be one of the most significant unit operations. The technology inherits certain advantages over other methods. These advantages include compactness and light weight, low labor intensity, modular design that allows for easy expansion or operation at partial capacity, low maintenance, low energy requirements, low cost, and environmentally friendly operations. A schematic representation of a simple separation membrane process is shown in Figure 19.16.

A feed stream of mixed components enters a membrane unit where it is separated into a retentate and permeate stream. The retentate stream is typically the purified product stream and the permeate stream contains the waste component.

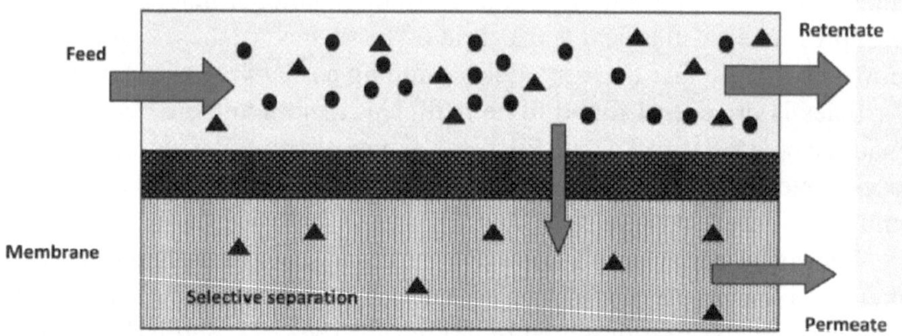

FIGURE 19.16 Schematic of membrane separation.

A quantitative measure of transport is the flux (or permeation rate), which is defined as the number of molecules that pass through a unit area/unit time [85]. It is believed that this molecular flux follows Fick's first law. The flux is proportional to the concentration gradient through the membrane. There is a movement from regions of high concentration to regions of low concentration, which may be expressed in the form:

$$J = -D\frac{dc}{dx} \tag{19.5}$$

By assuming a linear concentration gradient across the membrane, the flux can be approximated as:

$$J = -D\frac{C_2 - C_1}{L} \tag{19.6}$$

Where $C_1 = c(0)$ and $C_2 = c(L)$ are the downstream and upstream concentrations (corresponding to the pressures p_1 and p_2 via sorption isotherm c(p), respectively, and L is the membrane thickness, as labeled in Figure 19.17.

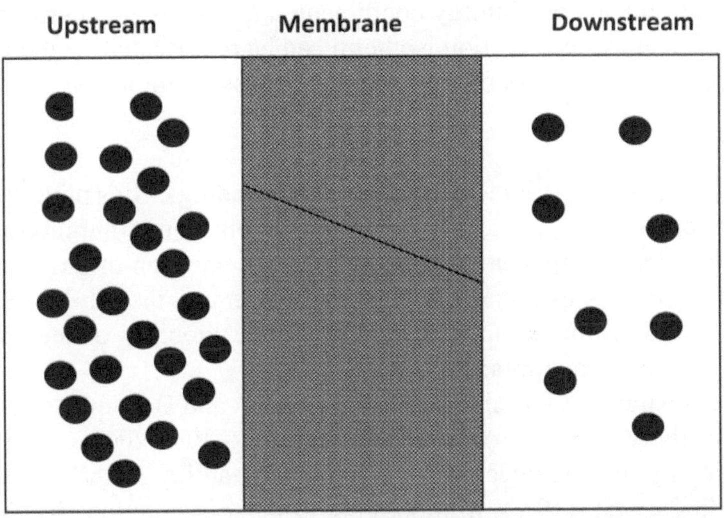

FIGURE 19.17 Separation membrane with a constant concentration gradient across membrane thickness L.

The membrane performance of various materials is commonly compared using the thickness, independent material property, the permeability, which is related to the flux as:

$$P = \frac{JL}{P_2 - P_1} = \left(\frac{C_2 - C_1}{P_2 - P_1}\right)D \tag{19.7}$$

In the case where the upstream pressure is much greater than the downstream pressure ($p_2 >> p_1$ and $C_2 >> C_1$) the permeability can be simplified so:

$$P = \frac{C_2}{P_2} D \tag{19.8}$$

The permeability is more commonly used to describe the performance of a membrane than flux. This is because the permeability of a homogenous stable membrane material is constant regardless of the pressure differential or membrane thickness, and hence it is easier to compare membranes made from different materials.

By introducing a solubility coefficient, the ratio of concentration over pressure C_2/p_2, when sorption isotherm can be represented by the Henry's law, the permeability coefficient may be expressed simply as:

$$P = SD \tag{19.9}$$

This form is useful as it facilitates the understanding of this physical property by representing it in terms of two components, one solubility which is an equilibrium component describing the concentration of gas molecules within the membrane,, and is the driving force, and the other diffusivity, which is a dynamic component describing the mobility of the gas molecules within the membrane.

The separation of a mixture of molecules A and B is characterized by the selectivity or ideal separation factor $\alpha_{A/B}$ = P (A)/P(B), the ratio of permeability of the molecule A over the permeability of the molecule B. According to Eq. (19.9), it is possible to make separations by diffusivity selectivity D(A)/D(B) or solubility selectivity S(A)/S(B) [85–86]. This formalism is known in membrane science as the solution-diffusion mechanism. Since the limiting stage of the mass transfer is overcoming of the diffusion energy barrier, this mechanism implies the activated diffusion. Based on this, the temperature dependences of the diffusion coefficients and permeability coefficients are described by the Arrhenius equations.

Gas molecules that encounter geometric constrictions experience an energy barrier such that sufficient kinetic energy of the diffusing molecule or the groups that form this barrier, in the membrane is required in order to overcome the barrier and make a successful diffusive jump. The common form of the Arrhenius dependence for the diffusion coefficient can be expressed as:

$$D_A = D_A^* \exp(-\Delta E_a / RT) \qquad (19.10)$$

For the solubility coefficient the van't Hoff equation holds:

$$S_A = S_A^* \exp(-\Delta H_a / RT) \qquad (19.11)$$

Where $\Delta H_a < 0$ is the enthalpy of sorption. From Eq. (19.9), it can be written:

$$P_A = P_A^* \exp(-\Delta E_P / RT) \qquad (19.12)$$

Where $\Delta E_p = \Delta E_a + \Delta H_a$ are known to diffuse within nonporous or porous membranes according to various transport mechanisms. Table 19.3 illustrates the mechanism of transport depending on the size of pores. For very narrow pores, size sieving mechanism is realized and that can be considered as a case of activated diffusion. This mechanism of diffusion is most common in the case of extensively studied nonporous polymeric membranes. For wider pores, the surface diffusion (also an activated diffusion process) and the Knudsen diffusion are observed [87–89].

TABLE 19.3 Transport mechanisms

Mechanism	Schamatic	Process
Activated diffusion	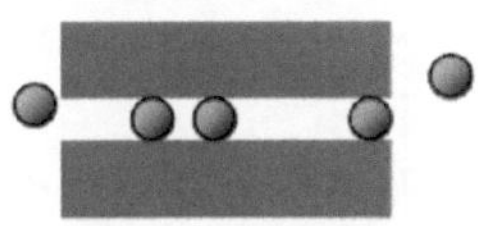	Constriction energy barrier 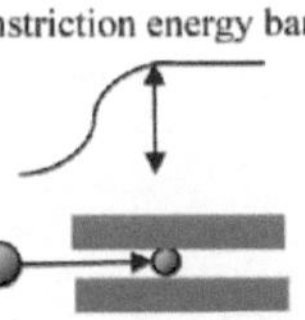
Surface diffusion	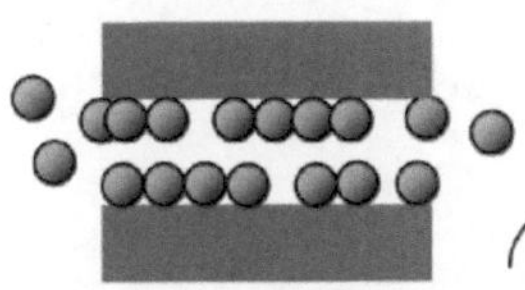	Adsorption – site energy barrier 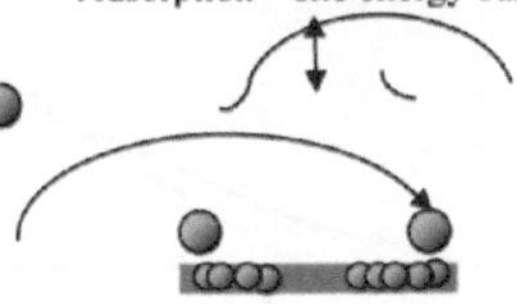
Knudsen diffusion	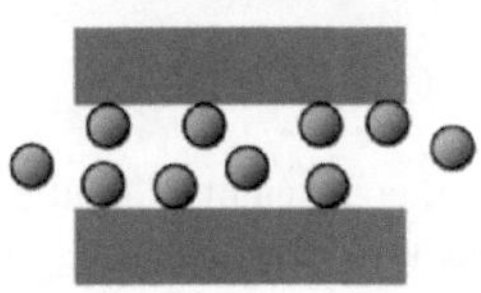	Direction and velocity 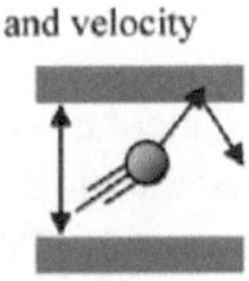

Sorption does not necessarily follow Henry's law. For a glassy polymer an assumption is made that there are small cavities in the polymer and the sorption at the cavities follows Langmuir's law. Then, the concentration in the membrane is given as the sum of Henry's law of adsorption and Langmuir's law of adsorption.

$$C = K_p P + \frac{C_h^* b_p}{1 + b_p} \tag{19.13}$$

It should be noted that the applicability of solution (sorption)–diffusion model has nothing to do with the presence or absence of the pore.

19.5.2 DIFFUSIVITY

The diffusivity through membranes can be calculated using the time-lag method [90]. A plot of the flow through the membrane versus time reveals an initial transient permeation followed by steady state permeation. Extending the linear section of the plot back to the intersection of the x-axis gives the value of the time-lag (θ) as shown in Figure 19.18.

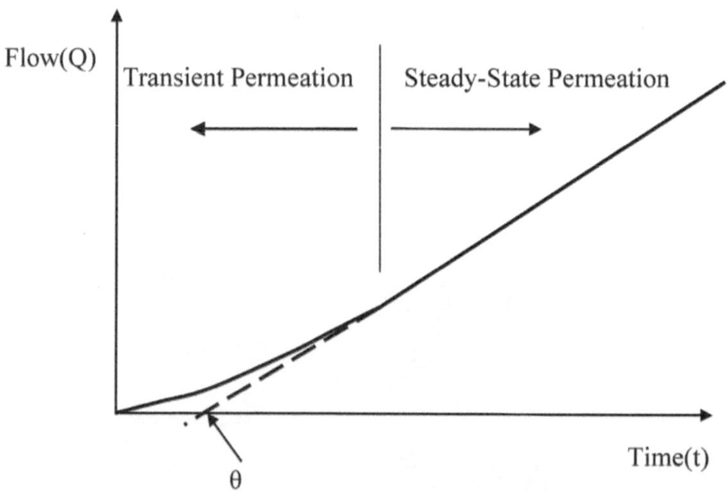

FIGURE 19.18 Calculation of the diffusion coefficient using the time-lag method, once the gradient is constant and steady state flow through the membrane has been reached, an extrapolation of the steady state flow line back to the x-axis where the flow is 0 reveals the value of the time lag (θ).

The time lag relates to the time it takes for the first molecules to travel through the membrane and is thus related to the diffusivity. The diffusion coefficient can be calculated from the time-lag and the membrane thickness as shown in Eq. (19.14) [91–92].

$$D = \frac{\Delta x}{6\theta} \tag{19.14}$$

Surface diffusion is the diffusion mechanism which dominates in the pore size region between activation diffusion and Knudsen diffusion [93].

19.5.3 SURFACE DIFFUSION

A model that describes well the surface diffusion on the pore walls was proposed many years ago. It was shown to be consistent with transport parameters in porous polymeric membranes. When the pore size decreased below a certain level, which depends on both membrane material and the permeability coefficient, the value of which exceeds for free molecular flow (Knudsen diffusion), especially in the case of organic vapors. Note that surface diffusion usually occurs simultaneously with Knudsen diffusion but it is the dominant mechanism within a certain pore size. Since surface diffusion is also a form of activated diffusion, the energy barrier is the energy required for the molecule to jump from one adsorption site to another across the surface of the pore. By allowing the energy barrier to be proportionate to the enthalpy of adsorption, Gilliland et al. established an equation for the surface diffusion coefficient expressed here as: [94]

$$D_S = D_S^* \exp(\frac{-aq}{RT}) \tag{19.15}$$

Where is a preexponential factor depending on the frequency of vibration of the adsorbed molecule normal to the surface and the distance from one adsorption site to the next. The quantity the heat of adsorption is (q > 0), and a proportionality constant is (0 < a < 1). The energy barrier separates the adjacent adsorption sites. An important observation is that more strongly adsorbed molecules are less mobile than weakly adsorbed molecules [95].

In the case of surface diffusion, the concentration is well described by Henry's law c = Kp, where K is $K = K_0 \exp(q/RT)$ [95–96]. Since solubility is the ratio of the equilibrium concentration over pressure, the solubility is equivalent to the Henry's law coefficient.

$$S_S = K_0 \exp\left(q|RT\right) \tag{19.16}$$

Which implies the solubility is a decreasing function of temperature. The product of diffusivity and solubility gives:

$$P_S = P_S^* \exp\left(\frac{(1-a)q}{RT}\right) \tag{19.17}$$

Since 0 < a < 1 the total permeability will decrease with increased temperature meaning that any increase in the diffusivity is counteracted by a decrease in surface concentration [95].

19.5.4 KNUDSEN DIFFUSION

Knudsen diffusion [95, 97–99] depends on pressure and mean free path which applies to pores between 10 Å and 500 Å in size [100]. In this region, the mean free path of molecules is much larger than the pore diameter. It is common to use Knudsen number $K_n = \lambda/d$ to characterize the regime of permeation through pores. When $K_n \ll 1$, viscous (Poiseuille) flow is realized. The condition for Knudsen diffusion is $Kn \gg 1$. An intermediate regime is realized when $K_n \approx 1$. The Knudsen diffusion coefficient can be expressed in the following form:

$$D_K = \frac{d}{3\tau}\bar{u} \tag{19.18}$$

This expression shows that the separation outcome should depend on the differences in molecular speed (or molecular mass). The average molecular speed is calculated using the Maxwell speed distribution as:

$$\bar{u} = \sqrt{\frac{8RT}{\pi m}} \tag{19.19}$$

And the diffusion coefficient can be presented as:

$$D_K = \left(\frac{d}{3\tau}\right)\left(\frac{8RT}{\pi m}\right)^{1/2} \tag{19.20}$$

For the flux in the Knudsen regime the following equation holds [101–102]:

$$J = n\pi d^2 \Delta p D_K / 4RTL \tag{19.21}$$

After substituting Eq. (19.20) into Eq. (19.21), one has the following expressions for the flux and permeability coefficient is:

$$J = \left(\frac{n\pi^{\frac{1}{2}} d^3 \Delta p}{6\tau L}\right)\left(\frac{2}{mRT}\right)^{1/2} \tag{19.22}$$

$$P = \left(\frac{n\pi^{\frac{1}{2}} d^3}{6\tau}\right)\left(\frac{2}{mRT}\right)^{1/2} \tag{19.23}$$

Two important conclusions can be made from analysis of Eqs. (19.22) and (19.23). First, selectivity of separation in Knudsen regime is characterized by the ratio $\alpha_{ij} = (M_j/M_i)^{1/2}$. It means that membranes where Knudsen diffusion predominates are poorly selective.

The most common approach to obtain diffusion coefficients is equilibrium molecular dynamics. The diffusion coefficient that is obtained is a self-diffusion coefficient. Transport-related diffusion coefficients are less frequently studied by simulation but several approaches using non-equilibrium MD (NEMD) simulation can be used.

19.5.5 MOLECULAR DYNAMICS (MD) SIMULATIONS

Conducting experiments for material characterization of the nanocomposites is a very time consuming, expensive, and difficult. Many researchers are now concentrating on developing both analytical and computational simulations. Molecular Dynamics simulations are widely being used in modeling and solving problems based on quantum mechanics. Using molecular dynamics it is possible to study the reactions, load transfer between atoms and molecules. If the objective of the simulation is to study the overall behavior of CNT-based composites and structures, such as deformations, load and heat transfer mechanisms then the continuum mechanics approach can be applied safely to study the problem effectively [103].

MD tracks the temporal evolution of a microscopic model system by integrating the equations of motion for all microscopic degrees of freedom. Numerical integration algorithms for initial value problems are used for this purpose, and their strengths and weaknesses have been discussed in simulation texts [104–106].

MD is a computational technique in which a time evolution of a set of interacting atoms is followed by integrating their equations of motion. The forces between atoms are because of the interactions with the other atoms. A trajectory is calculated in a 6-N dimensional phase space (three position and three momentum components for each of the N atoms). Typical MD simulations of CNT composites are performed on molecular systems containing up to tens to thousands of atoms and for simulation times up to nanoseconds. The physical quantities of the system are represented by averages over configurations distributed according to the chosen statistical ensemble. A trajectory obtained with MD provides such a set of configurations. Therefore the computation of a physical quantity is obtained as an arithmetic average of the instantaneous values. Statistical mechanics is the link between the nanometer behavior and thermodynamics. Thus the atomic system is expected to behave differently for different pressures and temperatures [107].

The interactions of the particular atom types are described by the total potential energy of the system, U, as a function of the positions of the individual atoms at a particular instant in time

$$U = U\left(X_i, \ldots\ldots, X_n\right) \tag{19.24}$$

where $_i$ represents the coordinates of atom i in a system of N atoms. The potential equation is invariant to the coordinate transformations, and is expressed in terms of the relative positions of the atoms with respect to each other, rather than from absolute coordinates [107].

MD is readily applicable to a wide range of models, with and without constraints. It has been extended from the original microcanonical ensemble formulation to a variety of statistical mechanical ensembles. It is flexible and valuable for extracting dynamical information. The Achilles "heel of MD" is its high demand of computer time, as a result of which the longest times that can be simulated with MD fall short of the longest relaxation times of most real-life macromolecular systems by several orders of magnitude. This has two important consequences: (1) Equilibrating an atomistic model polymer system with MD alone is problematic; if one starts from an improbable configuration, the simulation will not have the time to depart significantly from that configuration and visit the regions of phase space that contribute most significantly to the properties and (2) Dynamical processes with characteristic times longer than approximately 10^{-7} sec cannot be probed directly; the relevant correlation functions do not decay to zero within the simulation time and thus their long-time tails are inaccessible, unless some extrapolation is invoked based on their short-time behavior.

Recently, rigorous multiple time step algorithms have been invented, which can significantly augment the ratio of simulated time to CPU time. Such an algorithm is the reversible Reference System Propagator Algorithm (rRESPA) [108–109]. This algorithm invokes a Trotter factorization of the Liouville operator in the numerical integration of the equations of motion such that the fast-varying (e.g., bond stretching and bond angle bending) forces are updated with a short time step , while slowly varying forces (e.g., nonbonded interactions), which are typically expensive to calculate, are updated with a longer time step . Using and , one can simulate 300 ns of real time of a polyethylene melt on a modest workstation [110]. This is sufficient for the full relaxation of a system of C_{250} chains, but not of longer-chain systems.

A paper of Furukawa and Nitta is cited first to understand the NEMD simulation semiquantitatively, since, even though the paper deals with various pore shapes, the complicated simulation procedure is described clearly.

MD simulation is more preferable to study the nonequilibrium transport properties. Recently some NEMD methods have also been developed, such as the grand canonical molecular dynamics (GCMD) method [111–112] and the dual control volume GCMD technique (DCV- GCMD) [113–114]. These methods provide a valuable clue to insight into the transport and separation of fluids through a porous medium. The GCMD method has recently been used to investigate pressure-driven and chemical potential-driven gas transport through porous inorganic membrane [115].

19.5.5.1 EQUILIBRIUM MD SIMULATION

A self-diffusion coefficient can be obtained from the mean-square displacement (MSD) of one molecule by means of the Einstein equation in the form [115]:

$$D_A^* = \frac{1}{6N_\alpha} \lim_{t\to\infty} \lim_{t\to\infty} \frac{d}{dt}\left(r_i(t) - r_i(0)\right)^2 \tag{19.25}$$

Where Nα is the number of molecules, and are the initial and final (at time t) positions of the center of mass of one molecule i over the time interval t, and is MSD averaged over the ensemble. The Einstein relationship assumes a random walk for the diffusing species. For slow diffusing species, anomalous diffusion is sometimes observed and is characterized by:

$$\left(r_i(t) - r_i(0)\right)^2 \propto t^n \tag{19.26}$$

Where n < 1 (n = 1 for the Einstein diffusion regime). At very short times (t < 1 ps), the MSD may be quadratic iv n time (n = 2) which is characteristic of "free flight" as may occur in a pore or solvent cage prior to collision with the pore or cage wall. The result of anomalous diffusion, which may or may not occur in intermediate time scales, is to create a smaller slope at short times, resulting in a larger value for the diffusion coefficient. At sufficiently long times (the hydrodynamic limit), a transition from anomalous to Einstein diffusion (n = 1) may be observed [71].

An alternative approach to MSD analysis makes use of the center-of-mass velocity autocorrelation function (VACF) or Green–Kubo relation, given as follows [116]:

$$D = \frac{1}{3}\int (v_i(t).v_i(0))dt \tag{19.27}$$

Concentration in the simulation cell is extremely low and its diffusion coefficient is an order of magnitude larger than that of the polymeric segments. Under these circumstances, the self-diffusion and mutual diffusion coefficients of the penetrant are approximately equal, as related by the Darken equation in the following form:

$$D_{AB} = (D_A^* x_B + D_B^* x_A)\left(\frac{d \ln \ln f_A}{d\ lnc_A}\right) \tag{19.28}$$

In the limit of low concentration of diffusion , Eq. (19.28) reduces to:

$$D_A^* \equiv D_{AB} \tag{19.29}$$

19.5.5.2 NON-EQUILIBRIUM MD SIMULATION

Experimental diffusion coefficients, as obtained from time-lag measurements, report a transport diffusion coefficient which cannot be obtained from equilibrium MD simulation. Comparisons made in the simulation literature are typically between time-lag diffusion coefficients (even calculated for glassy polymers without correction for dual-mode contributions) and self-diffusion coefficients. As discussed above, mutual diffusion coefficients can be obtained directly from equilibrium MD simulation but simulation of transport diffusion coefficients require the use of NEMD methods, that are less commonly available and more computationally expensive [117].

For these reasons, they have not been frequently used. One successful approach is to simulate a chemical potential gradient and combine MD with GCMC methods (GCMC–MD), as developed by Heffelfinger and coworkers [114] and MacElroy [118]. This approach has been used to simulate permeation of a variety of small molecules through nanoporous carbon membranes, carbon nanotubes, porous silica, and self-assembled monolayers [119–121]. A diffusion coefficient then can be obtained from the relation:

$$D = \frac{KT}{F}(V) \tag{19.30}$$

19.5.6 GRAND CANONICAL MONTE CARLO (GCMC) SIMULATION

A standard GCMC simulation is employed in the equilibrium study, while MD simulation is more preferable to study the nonequilibrium transport properties [104].

Monte Carlo method is formally defined by the following "Numerical methods that are known as Monte Carlo methods can be loosely described as statistical simulation methods, where statistical simulation is defined in quite general terms to be any method that utilizes sequences of random numbers to perform the simulation [122]."

The name "Monte Carlo" was chosen because of the extensive use of random numbers in the calculations [104]. One of the better known applications of Monte Carlo simulations consists of the evaluation of integrals by generating suitable random numbers that will fall within the area of integration. A simple example of how a MC simulation method is applied to evaluate the value of π is illustrated in Figure 19.19. By considering a square that inscribes a circle of a diameter R, one can deduce that the area of the square is R^2, and the circle has an area of $\pi R^2/4$. Thus, the relative area of the circle and the square will be π/4. A large number of two independent random numbers (with x and y coordinates) of trial shots is generated within the square to determine whether each of them falls inside of the circle or not. After thousands or millions of trial shots, the computer program keeps counting the total number of trial shots inside the square and the number of shots landing inside the circle. Finally, the value of π/4 can be approximated based on the ratio of the number of shots that fall inside the circle to the total number of trial shots.

As stated earlier, the value of an integral can be calculated via MC methods by generating a large number of random points in the domain of that integral. Equation (19.31) shows a definite integral:

$$F = \int_a^b f(x)\,dx \tag{19.31}$$

Where f(x) is a continuous and real-valued function in the interval [a, b]. The integral can be rewritten as [104]:

$$F = \int_a^b dx \left(\frac{f(x)}{\rho(x)} \right) \rho(x) \cong \frac{f(\xi_i)}{\rho(\xi_i)}_{\tau} \tag{19.32}$$

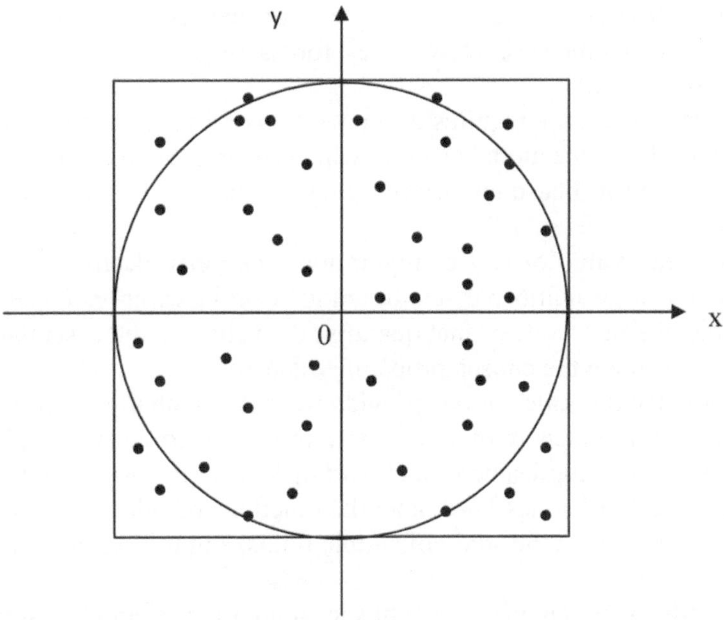

FIGURE 19.19 Illustration of the application of the Monte Carlo simulation method for the calculation of the value of π by generating a number of trial shots, in which the ratio of the number of shots inside the circle to the total number of trial shots will approximately approach the ratio of the area of the circle to the area of the square.

If the probability function is chosen to be a continuous uniform distribution, then:

$$\rho(x) = \frac{1}{(b-a)} a \le x \le b \tag{19.33}$$

Subsequently, the integral, F, can be approximated as:

$$F \approx \frac{(b-a)}{\tau} \sum^{i=1\tau} f(\xi_i) \tag{19.34}$$

In a similar way to the MC integration methods, MC molecular simulation methods rely on the fact that a physical system can be defined to possess a definite energy distribution function, which can be used to calculate thermodynamic properties.

The applications of MC are diverse such as nuclear reactor simulation, quantum chromodynamics, radiation cancer therapy, traffic flow, stellar evolution, econometrics, Dow Jones forecasting, oil well exploration, VSLI design [122].

The MC procedure requires the generation of a series of configurations of the particles of the model in a way which ensures that the configurations are distributed in phase space according to some prescribed probability density.

The mean value of any configurational property determined from a sufficiently large number of configurations provides an estimate of the ensemble-average value of that quantity; the nature of the ensemble average depends upon the chosen probability density.

These machine calculations provide what is essentially exact information on the consequences of a given intermolecular force law. Application has been made to hard spheres and hard disks, and to particles interacting through a Lennard-Jones 12–6 potential function and other continuous potentials of interest in the study of simple fluids, and to systems of charged particles [123].

The MC technique is a stochastic simulation method designed to generate a long sequence, or “Markov chain” of configurations that asymptotically sample the probability density of an equilibrium ensemble of statistical mechanics [105, 116]. For example, a MC simulation in the canonical (NVT) ensemble, carried out under the macroscopic constraints of a prescribed number of molecules N, total volume V, and temperature T, samples configurations r_P with probability proportional to , with, k_B being the Boltzmann constant, and T the absolute temperature. Thermodynamic properties are computed as averages over all sampled configurations.

The efficiency of a MC algorithm depends on the elementary moves it employs to go from one configuration to the next in the sequence. An attempted move typically involves changing a small number of degrees of freedom; it is accepted or rejected according to selection criteria designed so that the sequence ultimately conforms to the probability distribution of interest. In addition to usual moves of molecule translation and rotation practiced for small-molecule fluids, special moves have been invented for polymers. The reptation (slithering snake) move for polymer chains involves deleting a terminal segment on one end of the chain and appending a terminal segment on the other end, with the newly created torsion angle being assigned a randomly chosen value [124].

In most MC algorithms the overall probability of transition from some state (configuration) m to some other state n, as dictated by both the attempt and the selection stages of the moves, equals the overall probability of transition from n to m. This is the principle of detailed balance or "microscopic reversibility." The probability of attempting a move from state m to state n may or may not be equal to that of attempting the inverse move from state n to state m. These probabilities of attempt are typically unequal in "bias" MC algorithms, which incorporate information about the system energetics in attempting moves. In bias MC, detailed balance is ensured by appropriate design of the selection criterion, which must remove the bias inherent in the attempt [105, 116].

19.5.7 MEMBRANE MODEL AND SIMULATION BOX

The MD simulations [125] can be applied for the permeation of pure and mixed gasses across carbon membranes with three different pore shapes: the diamond pore (DP), zigzag path (ZP) and straight path (SP), each composed of micro-graphite crystalline. Three different pore shapes can be considered: DP, ZP, and SP.

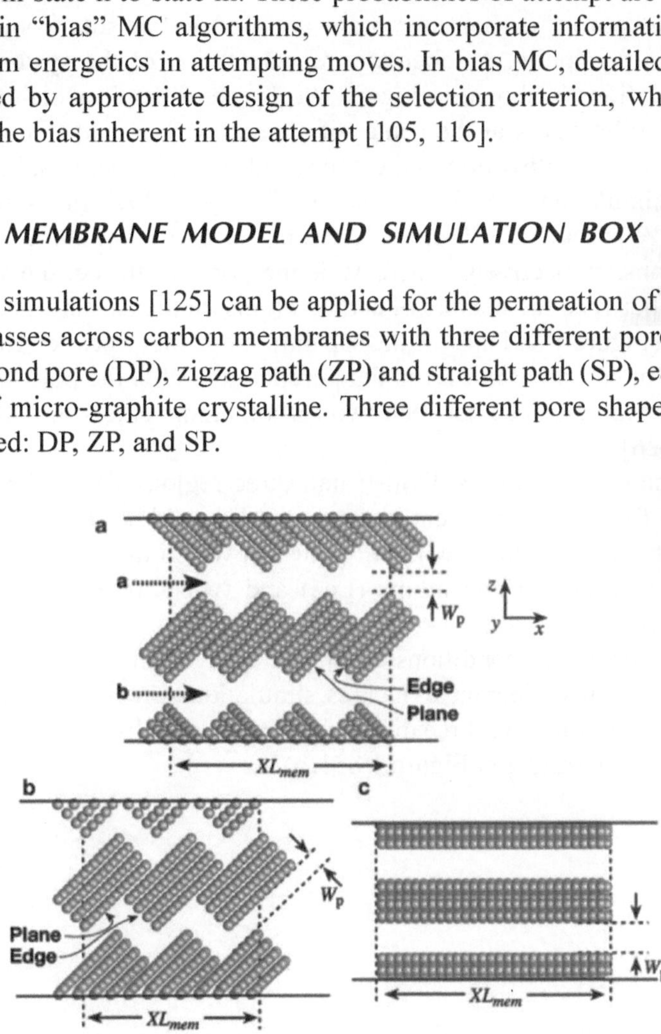

FIGURE 19.20 Three membrane pore shapes; (a) diamond path (DP), (b) zigzag path (ZP), (c) straight path (SP).

Figure 19.20 A–C shows the cross-sectional view of each pore shape. DP (A) has two different pore mouths; one a large (pore a) and the other a small mouth (pore b). ZP (B) has zigzag shaped pores whose sizes (diameters) are all the same at the pore entry. SP (C) has straight pores which can be called slit-shaped pores.

In a simulation system, we investigate the equilibrium of selective adsorption and nonequilibrium transport, and the separation of gas mixture in the nanoporous carbon membrane are modeled as slits from the layer structure of graphite. A schematic representation of the system used in our simulations is shown in Figure 19.21(a) and (b), in which the origin of the coordinates is at the center of simulation box and transport takes place along the x-direction in the nonequilibrium simulations. In the equilibrium simulations, the box as shown in Figure 19.21(a) is employed, whose size is set as 85.20 nm × 4.92 nm × (1.675 + W) nm in x-, y-, and z-directions, respectively, where W is the pore width, i.e. the separation distance between the centers of carbon atoms on the two layers forming a slit pore (Figure 19.21). L_{cc} is the separation distance between two centers of adjacent carbon atom; L_m is the pore length; W is the pore width, Δ is the separation distance between two carbon atom centers of two adjacent layers [126].

The simulation box is divided into three regions where the chemical potential for each component is the same. The middle region (M-region) represents the membrane with slit pores. In which the distances between the two adjacent carbon atoms (Lcc) and two adjacent graphite basal planes (Δ).

Period boundary conditions are employed in all three directions. In the nonequilibrium molecular dynamics simulations, in order to use period boundary conditions in three directions, we have to divide the system into five regions as shown in Figure 19.21(b).

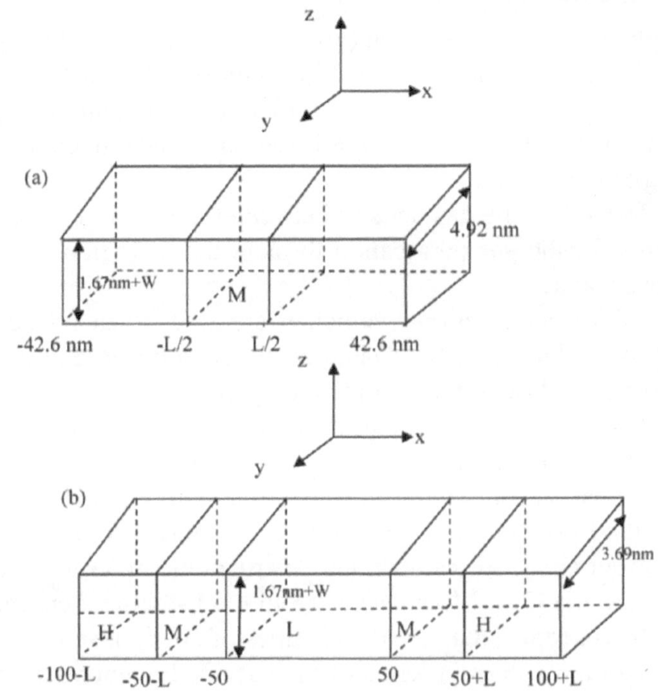

FIGURE 19.21 Schematic representation of the simulation boxes. The H-, L- and M-areas correspond to the high and low chemical potential control volumes, and membrane, respectively. Transport takes place along the x-direction in the nonequilibrium simulations. (a) Equilibrium adsorption simulations and (b) nonequilibrium transport simulations. L is the membrane thickness and W is the pore width (Figure 19.22).

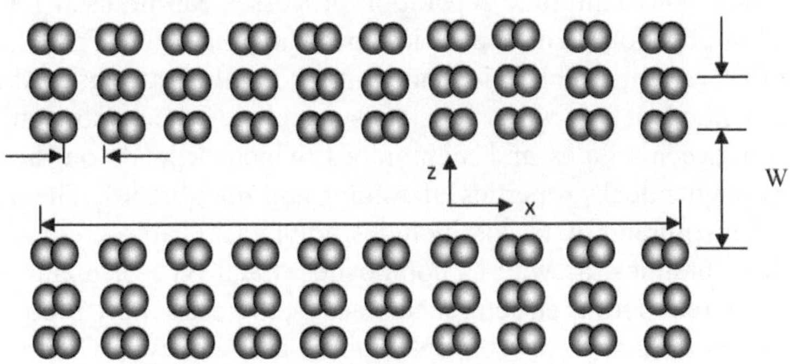

FIGURE 19.22 Schematic representation of slit pore.

Each symmetric box has three regions. Two are density control; H-region (high density) and L-region (low density) and one is free of control that is the M-region which is placed between the H- and L- region. For each simulation, the density in the H-region, ρ_H, is maintained to be that of the feed gas and the density in the L-region is maintained at zero, corresponding to the vacuum.

The difference in the gas density between the H- and L- region is the driving force for the gas permeation through the M-region which represents the membrane.

The transition and rotational velocities are given to each inserted molecules randomly based on the Gaussian distribution around an average velocity corresponding to the specified temperature.

Molecules spontaneously move from H- to L-region via leap-frog algorithm and a nonequilibrium steady state is obtained at the M-region. During a simulation run, equilibrium with the bulk mass at the feed side at the specified pressure and temperature is maintained at the H-region by carrying out GCMC creations and destructions, in terms of the usual acceptance criteria [28]. Molecules entered the L-region were moved out immediately to keep vacuum. The velocities of newly inserted molecules were set to certain values in terms of the specified temperature by use of random numbers on the Gaussian distribution.

19.6 CONCLUDING REMARKS

The concept of membrane processes is relatively simple but nevertheless often unknown. Membrane separation processes can be used for a wide range of applications. The separation mechanism in MF/UF/NF is mainly the size exclusion, which is indicated in the nominal ratings of the membranes. The other separation mechanism includes the electrostatic interactions between solutes and membranes, which depends on the surface and physiochemical properties of solutes and membranes. The available range of membrane materials includes polymeric, carbon, silica, zeolite and other ceramics, as well as composites. Each type of membrane can have a different porous structure. Nowadays, there are more reports on the fluid transport through porous CNTs/polymer membrane. Computational approach can play an important role in the development of the CNT-based composites by providing simulation results to help with the understanding, analysis, and design of such nanocomposites. Computational approaches to obtain solubility and diffusion coefficients of small molecules in polymers are focused primarily upon molecular dynamics and Monte Carlo methods. Molecular dynamics simulations are widely being used in modeling and solving problems based on quantum mechanics. Using molecular dynamics it is possible to study the reactions, and load transfer between atoms and molecules. Monte Carlo molecular simulation methods rely on the fact that a physical system can be defined to possess a definite energy distribution function, which can be used to calculate thermodynamic properties. The Monte Carlo technique is a stochastic simulation method designed to generate a long sequence, or "Markov chain" of configurations that asymptotically sample the probability density of an equilibrium ensemble of statistical mechanics. So using the molecular dynamic or Monte Carlo techniques can be useful to simulate the membrane separation process which in turn depends on the purpose and the condition of process.

SYMBOLS

Symbol	**Definition**
r_p	Pore radius
Γ	Surface tension
T	Thickness of the adsorbate film
V_L	Molecular volume of the condensate
a_1 *and* a_2	Unit vectors
G_1, G_2	The material coordinates of a point in the initial configuration
e_1, e_2 *and* e_3	The coordinates in the current configuration
R	The radius of the modeled SWCNT
J	Molecular flux
D	The diffusivity (diffusion coefficient)
$c\,(x)$	Concentration
X	Position across the membrane
P	Permeability
S	Solubility coefficient
K_p	Henry's law constant
B	Hole affinity constant
C_h^*	Saturation constant

Q	Flow
T	Time
Θ	Time-lag
Q	Heat of adsorption
A	Proportionality constant
aq	Energy barrier
K	Temperature dependent Henry's law coefficient
K_0	Proportionality constant
P_S^*	Constant
Λ	The mean free path of molecules
D	Pore diameter
K_n	Knudsen number
T	Pore tortuosity
$\overline{u}$	Average molecular speed
M	Molecular mass
N	Surface concentration of pores
Δp	Pressure drop across the membrane
L	Membrane thickness
D_A^*	Self-diffusion coefficient
Na	The number of molecules
f_A	Fugacity
c_A	The concentration of diffusant A
F	Applied external force
V	The center-of-mass velocity component
$\rho(x)$	An arbitrary probability distribution function
ξ_i	The random numbers generated for each trial
T	The number of trials
RO	Reverse osmosis
NF	Nanofiltration
UF	Ultrafiltration
MF	Microfiltration
MWCO	Molecular Weight Cut-Off
IUPAC	International Union of Pure and Applied Chemistry
SWNT	Single-Walled Carbon Nanotube
MWNT	Multi-Walled Carbon Nanotube

CNT	Carbon nanotube
AFM	Atomic Force Microscopy
MD	Molecular Dynamics
MC	Monte Carlo
NEMD	Non-Equilibrium MD
RESPA	Reference System Propagator Algorithm
GCMD	Grand Canonical Molecular Dynamics
DCV- GCMD	Dual-volume GCMD
MSD	Mean-Square Displacement
DP	Diamond Pore
ZP	Zigzag Path
SP	Straight Path

KEYWORDS

- **Computational methods**
- **Filtration**
- **Membrane**
- **Membrane types**

REFERENCES

1. Majeed, S.; et al. Multi-walled carbon nanotubes (MWCNTs) mixed polyacrylonitrile (PAN) ultrafiltration membranes. *J. Membr. Sci.* **2012,** *403,* 101–109.
2. Macedonio, F.; and Drioli, E.; Pressure-driven membrane operations and membrane distillation technology integration for water purification. *Desalination.* **2008,** *223(1),* 396–409.
3. Merdaw, A. A.; Sharif, A. O.; and Derwish, G. A. W.; Mass transfer in pressure-driven membrane separation processes, Part II. *Chem. Eng. J.* **2011,** *168(1),* 229–240.
4. Van Der Bruggen, B.; et al. A review of pressure-driven membrane processes in wastewater treatment and drinking water production. *Environ. Prog.* **2003,** *22(1),* 46–56.
5. Cui, Z. F.; and Muralidhara, H. S.; Membrane Technology: A Practical Guide to Membrane Technology and Applications in Food and Bioprocessing. Elsevier; **2010,** 288 p.
6. Shirazi, S.; Lin, C. J.; and Chen, D.; Inorganic fouling of pressure-driven membrane processes —A critical review. *Desalination.* **2010,** *250(1),* 236–248.
7. Pendergast, M. M.; and Hoek, E. M. V.; A review of water treatment membrane nanotechnologies. *Energ. Environ. Sci.* **2011,** *4(6),* 1946–1971.

8. Hilal, N.; et al. A comprehensive review of nanofiltration membranes: Treatment, pretreatment, modelling, and atomic force microscopy. *Desalination.* **2004,** *170(3),* 281–308.
9. Srivastava, A.; Srivastava, S.; and Kalaga, K.; Carbon Nanotube Membrane Filters, in Springer Handbook of Nanomaterials. Springer; **2013,** 1099–1116 pp.
10. Colombo, L.; and Fasolino, A. L.; Computer-Based Modeling of Novel Carbon Systems and Their Properties: Beyond Nanotubes. Springer; **2010,** *3,* 258 p.
11. Polarz, S.; and Smarsly, B.; Nanoporous materials. *J. Nanosci. Nanotechnol.* **2002,** *2(6),* 581–612.
12. Gray-Weale, A. A.; et al. Transition-state theory model for the diffusion coefficients of small penetrants in glassy polymers. Macromolecules. **1997,** *30(23),* 7296–7306.
13. Rigby, D.; and R. Roe, Molecular dynamics simulation of polymer liquid and glass. I. Glass transition. *J. Chem. Phys.* **1987,** *87,* 7285.
14. Freeman, B. D.; Yampolskii, Y. P.; and Pinnau, I.; Materials Science of Membranes for Gas and Vapor Separation. Wiley. com. **2006,** 466 p.
15. Hofmann, D.; et al. Molecular modeling investigation of free volume distributions in stiff chain polymers with conventional and ultrahigh free volume: Comparison between molecular modeling and positron lifetime studies. *Macromolecules.* **2003,** *36(22),* 8528–8538.
16. Greenfield, M. L.; and Theodorou, D. N.; Geometric analysis of diffusion pathways in glassy and melt atactic polypropylene. *Macromolecules.* **1993,** *26(20),* 5461–5472.
17. Baker, R. W.; Membrane Technology and Applications. John Wiley & Sons; **2012,** 592 p.
18. Strathmann, H.; Giorno, L.; and Drioli, E.; Introduction to Membrane Science and Technology. Wiley-VCH Verlag & Company; **2011,** 544 p.
19. Chen, J. P.; et al. Membrane Separation: Basics and Applications, in Membrane and Desalination Technologies. Wang, L. K.; et al. ed. Humana Press; **2008,** 271–332.
20. Mortazavi, S.; Application of membrane separation technology to mitigation of mine effluent and acidic drainage. Natural Resources Canada; **2008,** 194 p.
21. Porter, M. C.; Handbook of Industrial Membrane Technology. Noyes Publications; **1990,** 604 p.
22. Naylor, T. V.; Polymer Membranes: Materials, Structures and Separation Performance. Rapra Technology Limited; **1996,** 136 p.
23. Freeman, B. D.; Introduction to membrane science and technology. By Heinrich Strathmann. *Angewandte Chem. Int. Ed.* **2012,** *51(38),* 9485–9485.
24. Kim, I.; Yoon, H.; and Lee, K. M.; Formation of integrally skinned asymmetric polyetherimide nanofiltration membranes by phase inversion process. *J. Appl. Polym. Sci.* **2002,** *84(6),* 1300–1307.
25. Khulbe, K. C.; Feng, C. Y.; and Matsuura, T.; Synthetic Polymeric Membranes: Characterization by Atomic Force Microscopy. Springer; **2007,** 198 p.
26. Loeb, L. B.; The Kinetic Theory of Gases. Courier Dover Publications; **2004,** 678 p.
27. Koros, W. J.; and Fleming, G. K.; Membrane-based gas separation. *J. Membr. Sci.* **1993,** *83(1),* 1–80.
28. Perry, J. D.; Nagai, K.; and Koros, W. J.; Polymer membranes for hydrogen separations. *MRS Bull.* **2006,** *31(10),* 745–749.

29. Yang, W.; et al. Carbon nanotubes for biological and biomedical applications. *Nanotechnology.* **2007,** *18(41),* 412001.
30. Bianco, A.; et al. Biomedical applications of functionalised carbon nanotubes. *Chem. Commun.* **2005,** *5,* 571–577.
31. Salvetat, J.; et al. Mechanical properties of carbon nanotubes. *Appl. Phys. A.* **1999,** *69(3),* 255–260.
32. Zhang, X.; et al. Ultrastrong, stiff, and lightweight carbon-nanotube fibers. *Adv. Mater.* **2007,** *19(23),* 4198–4201.
33. Arroyo, M.; and Belytschko, T.; Finite crystal elasticity of carbon nanotubes based on the exponential cauchy-born rule. *Phys. Rev. B.* **2004,** *69(11),* 115415.
34. Wang, J.; et al. Energy and mechanical properties of single-walled carbon nanotubes predicted using the higher order cauchy-born rule. *Phys. Rev. B.* **2006,** *73(11),* 115428.
35. Zhang, Y.; Single-Walled Carbon Nanotube Modelling Based on One- and Two-Dimensional Cosserat Continua. University of Nottingham; **2011**.
36. Wang, S.; Functionalization of Carbon Nanotubes: Characterization, Modeling and Composite Applications. Florida State University; **2006,** 193 p.
37. Lau, K.-T.; Gu, C.; and Hui, D.; A critical review on nanotube and nanotube/nanoclay related polymer composite materials. *Composites Part B: Engineering.* **2006,** *37(6),* 425-436.
38. Choi, W.; et al. Carbon nanotube-guided thermopower waves. *Mater. Today.* **2010,** *13(10),* 22–33.
39. Iijima, S.; Helical microtubules of graphitic carbon. *Nature.* **1991,** *354(6348),* 56–58.
40. Sholl, D. S.; and Johnson, J.; Making high-flux membranes with carbon nanotubes. *Science.* **2006,** *312(5776),* 1003–1004.
41. Zang, J.; et al. Self-diffusion of water and simple alcohols in single-walled aluminosilicate nanotubes. *ACS Nano.* **2009,** *3(6),* 1548–1556.
42. Talapatra, S.; Krungleviciute, V.; and Migone, A. D.; Higher coverage gas adsorption on the surface of carbon nanotubes: Evidence for a possible new phase in the second layer. *Phys. Rev. Lett.* **2002,** *89(24),* 246106.
43. Pujari, S.; et al. Orientation dynamics in multiwalled carbon nanotube dispersions under shear flow. *J. Chem. Phys.* **2009,** *130,* 214903.
44. Singh, S.; and Kruse, P.; Carbon nanotube surface science. *Int. J. Nanotechnol.* **2008,** *5(9),* 900–929.
45. Baker, R. W.; Future directions of membrane gas separation technology. *Indus. Eng. Chem. Res.* **2002,** *41(6),* 1393–1411.
46. Erucar, I.; and Keskin, S.; Screening metal–organic framework-based mixed-matrix membranes for CO2/CH4 separations. *Indus. Eng. Chem. Res.* **2011,** *50(22),* 12606–12616.
47. Bethune, D. S.; et al. Cobalt-catalysed growth of carbon nanotubes with single-atomic-layer walls. *Nature.* **1993,** *363,* 605–607.
48. Iijima, S.; and Ichihashi, T.; Single-shell carbon nanotubes of 1-nm diameter. *Nature.* **1993,** *363,* 603–605.
49. Treacy, M.; Ebbesen, T.; and Gibson, J.; Exceptionally High Young's Modulus Observed for Individual Carbon Nanotubes. **1996**.
50. Wong, E. W.; Sheehan, P. E.; and Lieber, C.; Nanobeam mechanics: Elasticity, strength, and toughness of nanorods and nanotubes. *Science.* **1997,** *277(5334),* 1971–1975.

51. Thostenson, E. T.; Li, C.; and Chou, T. W.; Nanocomposites in context. *Compos. Sci. Technol.* **2005,** *65(3),* 491–516.
52. Barski, M.; Kędziora, P.; and Chwał, M.; Carbon nanotube/polymer nanocomposites: A brief modeling overview. *Key Eng. Mater.* **2013,** *542,* 29–42.
53. Dresselhaus, M. S.; Dresselhaus, G.; and Eklund, P. C.; Science of Fullerenes and Carbon nanotubes:Ttheir Properties and Applications. Academic Press; **1996,** 965 p.
54. Yakobson, B.; and Smalley, R. E.; Some unusual new molecules—long, hollow fibers with tantalizing electronic and mechanical properties—have joined diamonds and graphite in the carbon family. *Am. Sci.* **1997,** *85,* 324–337.
55. Guo, Y.; and Guo, W.; Mechanical and electrostatic properties of carbon nanotubes under tensile loading and electric field. *J. Phys. D: Appl. Phys.* **2003,** *36(7),* 805.
56. Berger, C.; et al. Electronic confinement and coherence in patterned epitaxial graphene. *Science.* **2006,** *312(5777),* 1191–1196.
57. Song, K.; et al. Structural polymer-based carbon nanotube composite fibers: understanding the processing–structure–performance relationship. *Material.* **2013,** *6(6),* 2543–2577.
58. Park, O. K.; et al. Effect of surface treatment with potassium persulfate on dispersion stability of multi-walled carbon nanotubes. *Mater. Lett.* **2010,** *64(6),* 718–721.
59. Banerjee, S.; Hemraj-Benny, T.; and Wong, S. S.; Covalent surface chemistry of single-walled carbon nanotubes. *Adv. Mater.* **2005,** *17(1),* 17–29.
60. Balasubramanian, K.; and Burghard, M.; Chemically functionalized carbon nanotubes. *Small.* **2005,** *1(2),* 180–192.
61. Xu, Z. L.; and Alsalhy Qusay, F.; Polyethersulfone (PES) hollow fiber ultrafiltration membranes prepared by PES/non-solvent/NMP solution. *J. Membr. Sci.* **2004,** *233(1–2),* 101–111.
62. Chung, T. S.; Qin, J. J.; and Gu, J.; Effect of shear rate within the spinneret on morphology, separation performance and mechanical properties of ultrafiltration polyethersulfone hollow fiber membranes. *Chem. Eng. Sci.* **2000,** *55(6),* 1077–1091.
63. Choi, J. H.; Jegal, J.; and Kim, W. N.; Modification of performances of various membranes using MWNTs as a modifier. *Macromole. Symp.* **2007,** *249–250(1),* 610–617.
64. Wang, Z.; and Ma, J. The role of nonsolvent in-diffusion velocity in determining polymeric membrane morphology. *Desalination.* **2012,** *286(0),* 69–79.
65. Vilatela, J. J.; Khare, R.; and Windle, A. H.; The hierarchical structure and properties of multifunctional carbon nanotube fibre composites. *Carbon.* **2012,** *50(3),* 1227–1234.
66. Benavides, R. E.; Jana, S. C.; and Reneker, D. H.; Nanofibers from scalable gas jet process. *ACS Macro Lett.* **2012,** *1(8),* 1032–1036.
67. Gupta, V. B.; and Kothari, V. K.; Manufactured Fiber Technology. Springer; **1997,** 661 p.
68. Wang, T.; and Kumar, S.; Electrospinning of polyacrylonitrile nanofibers. *J. Appl. Polym. Sci.* **2006,** *102(2),* 1023–1029.
69. Song, K.; et al. Lubrication of poly (vinyl alcohol) chain orientation by carbon nanochips in composite tapes. *J. Appl. Polym. Sci.* **2013,** *127(4),* 2977–2982.
70. Theodorou, D. N.; Molecular simulations of sorption and diffusion in amorphous polymers. *Plast. Eng.-New York.* **1996,** *32,* 67–142.

71. Müller-Plathe, F.; Permeation of polymers—a computational approach. *Acta Polym.* **1994,** *45(4),* 259–293.
72. Liu, Y. J.; and Chen, X. L.; Evaluations of the effective material properties of carbon nanotube-based composites using a nanoscale representative volume element. *Mech. Mater.* **2003,** *35(1),* 69–81.
73. Gusev, A. A.; and Suter, U. W.; Dynamics of small molecules in dense polymers subject to thermal motion. *J. Chem. Phys.* **1993,** *99,* 2228.
74. Elliott, J. A.; Novel approaches to multiscale modelling in materials science. *Int. Mater. Rev.* **2011,** *56(4),* 207–225.
75. Greenfield, M. L.; and Theodorou, D. N.; Molecular modeling of methane diffusion in glassy atactic polypropylene via multidimensional transition state theory. *Macromolecules.* **1998,** *31(20),* 7068–7090.
76. Peng, F.; et al. Hybrid organic-inorganic membrane: Solving the tradeoff between permeability and selectivity. *Chem. Mater.* **2005,** *17(26),* 6790–6796.
77. Duke, M. C.; et al. Exposing the molecular sieving architecture of amorphous silica using positron annihilation spectroscopy. *Adv. Funct. Mater.* **2008,** *18(23),* 3818–3826.
78. Hedstrom, J. A.; et al. Pore morphologies in disordered nanoporousthin films. *Langmuir.* **2004,** *20(5),* 1535–1538.
79. Pujari, P. K.; et al. Study of pore structure in grafted polymer membranes using slow positron beam and small-angle x-ray scattering techniques. *Nuclear Instr. Methods Phys. Res. Sec. B: Beam Inter. Mater. Atoms.* **2007,** *254(2),* 278–282.
80. Wang, X. Y.; et al. Cavity size distributions in high free volume glassy polymers by molecular simulation. *Polymer.* **2004,** *45(11),* 3907–3912.
81. Skoulidas, A. I.; and Sholl, D. S.; Self-diffusion and transport diffusion of light gases in metal-organic framework materials assessed using molecular dynamics simulations. *J. Phys. Chem. B.* **2005,** *109(33),* 15760–15768.
82. Wang, X. Y.; et al. A molecular simulation study of cavity size distributions and diffusion in para and meta isomers. *Polymer.* **2005,** *46(21),* 9155–9161.
83. Zhou, J.; et al. Molecular dynamics simulation of diffusion of gases in pure and silica-filled poly (1-trimethylsilyl-1-propyne)[PTMSP]. *Polymer.* **2006,** *47(14),* 5206–5212.
84. Scholes, C. A.; Kentish, S. E.; and Stevens, G. W.; Carbon dioxide separation through polymeric membrane systems for flue gas applications. *Recent Patents Chem. Eng.* **2008,** *1(1),* 52–66.
85. Wijmans, J. G.; and Baker, R. W.; The solution-diffusion model: a unified approach to membrane permeation. *Mater. Sci. Membr. Gas Vapor Separat.* **2006,** 159–190.
86. Wijmans, J. G.; and Baker, R. W.; The solution-diffusion model: A review. *J. Membr. Sci.* **1995,** *107(1),* 1–21.
87. Way, J. D.; and Roberts, D. L.; Hollow fiber inorganic membranes for gas separations. *Separat. Sci. Technol.* **1992,** *27(1),* 29–41.
88. Rao, M. B.; and Sircar, S.; Performance and pore characterization of nanoporous carbon membranes for gas separation. *J. Membr. Sci.* **1996,** *110(1),* 109–118.
89. Merkel, T. C.; et al. Effect of nanoparticles on gas sorption and transport in poly (1-trimethylsilyl-1-propyne). *Macromolecules.* **2003,** *36(18),* 6844–6855.
90. Mulder, M.; Basic Principles of Membrane Technology Second Edition. Kluwer Academic Publication; **1996,** 564 p.

91. Wang, K.; Suda, H.; and Haraya, K.; Permeation time lag and the concentration dependence of the diffusion coefficient of CO2 in a carbon molecular sieve membrane. *Indus. Eng. Chem. Res.* **2001,** *40(13),* 2942–2946.
92. Webb, P. A.; and Orr, C.; Analytical Methods in Fine Particle Technology. Micromeritics Norcross, GA. **1997,** *55,* 301.
93. Pinnau, I.; et al. Long-term permeation properties of poly (1-trimethylsilyl-1-propyne) membranes in hydrocarbon—vapor environment. *J. Polym. Sci. Part B: Polym. Phys.* **1997,** *35(10),* 1483–1490.
94. Jean, Y. C.; characterizing free volumes and holes in polymers by positron annihilation spectroscopy. *Positron Spectr. Solids.* **1993,** 1.
95. Hagiwara, K.; et al. Studies on the free volume and the volume expansion behavior of amorphous polymers. *Radiat. Phys. Chem.* **2000,** *58(5),* 525–530.
96. Sugden, S.; Molecular volumes at absolute zero. Part II. Zero volumes and chemical composition. *J. Chem. Soc. (Resumed).* **1927,** 1786–1798.
97. Dlubek, G.; et al. Positron annihilation: A unique method for studying polymers. In: Macromolecular Symposia. Wiley Online Library; **2004.**
98. Golemme, G.; et al. NMR study of free volume in amorphous perfluorinated polymers: comparsion with other methods. *Polymer.* **2003,** *44(17),* 5039–5045.
99. Victor, J. G.; and Torkelson, J. M.; On measuring the distribution of local free volume in glassy polymers by photochromic and fluorescence techniques. *Macromolecules.* **1987,** *20(9),* 2241–2250.
100. Royal, J. S.; and Torkelson, J. M.; Photochromic and fluorescent probe studies in glassy polymer matrices. *Macromolecules.* **1992,** *25(18),* 4792–4796.
101. Yampolskii, Y. P.; et al. Study of high permeability polymers by means of the spin probe technique. *Polymer.* **1999,** *40(7),* 1745–1752.
102. Kobayashi, Y.; et al. Evaluation of polymer free volume by positron annihilation and gas diffusivity measurements. *Polymer.* **1994,** *35(5),* 925–928.
103. Huxtable, S. T.; et al. Interfacial heat flow in carbon nanotube suspensions. *Nat. Mater.* **2003,** *2(11),* 731–734.
104. Allen, M. P.; and Tildesley, D. J.; Computer Simulation of Liquids. Oxford University Press; **1989.**
105. Frenkel, D.; Smit, B.; and Ratner, M. A.; Understanding molecular simulation: From algorithms to applications. *Phys. Today.* **1997,** *50,* 66.
106. Rapaport, D. C.; The Art of Molecular Dynamics Simulation. Cambridge University Press; **2004,** 549 p.
107. Leach, A. R.; and Schomburg, D.; Molecular Modelling: Principles and Applications. Longman London; **1996.**
108. Martyna, G. J.; et al. Explicit reversible integrators for extended systems dynamics. *Mole. Phys.* **1996,** *87(5),* 1117–1157.
109. Tuckerman, M.; Berne, B. J.; and Martyna, G. J.; Reversible multiple time scale molecular dynamics. *J. Chem. Phys.* **1992,** *97(3),* 1990.
110. Harmandaris, V. A.; et al. Crossover from the rouse to the entangled polymer melt regime: signals from long, detailed atomistic molecular dynamics simulations, supported by rheological experiments. *Macromolecules.* **2003,** *36(4),* 1376–1387.

111. Firouzi, M.; Tsotsis, T. T.; and Sahimi, M.; Nonequilibrium molecular dynamics simulations of transport and separation of supercritical fluid mixtures in nanoporous membranes. I. Results for a single carbon nanopore. *J. Chem. Phys.* **2003,** *119,* 6810.
112. Shroll, R. M.; and Smith, D. E.; Molecular dynamics simulations in the grand canonical ensemble: application to clay mineral swelling. *J. Chem. Phys.* **1999,** *111,* 9025.
113. Firouzi, M.; et al. Molecular dynamics simulations of transport and separation of carbon dioxide–alkane mixtures in carbon nanopores. *J. Chem. Phys.* **2004,** *120,* 8172.
114. Heffelfinger, G. S.; and Van Swol, F.; Diffusion in Lennard-Jones fluids using dual control volume grand canonical molecular dynamics simulation (DCV-GCMD). *J. Chem. Phys.* **1994,** *100,* 7548.
115. Pant, P. K.; and Boyd, R. H.; Simulation of diffusion of small-molecule penetrants in polymers. *Macromolecules.* **1992,** *25(1),* 494–495.
116. Allen, M. P.; and Tildesley, D. J.; Computer Simulation of Liquids. Oxford University Press; **1989,** 385 p.
117. Cummings, P. T.; and Evans, D. J.; Nonequilibrium molecular dynamics approaches to transport properties and Non-Newtonian fluid rheology. *Indus. Eng. Chem. Res.* **1992,** *31(5),* 1237–1252.
118. MacElroy, J.; Nonequilibrium molecular dynamics simulation of diffusion and flow in thin microporous membranes. *J. Chem. Phys.* **1994,** *101,* 5274.
119. Furukawa, S.; and Nitta, T.; Non-equilibrium molecular dynamics simulation studies on gas permeation across carbon membranes with different pore shape composed of micro-graphite crystallites. *J. Membr. Sci.* **2000,** *178(1),* 107–119.
120. Düren, T.; Keil, F. J.; and Seaton, N. A.; Composition dependent transport diffusion coefficients of CH₄/CF₄ mixtures in carbon nanotubes by non-equilibrium molecular dynamics simulations. *Chem. Eng. Sci.* **2002,** *57(8),* 1343–1354.
121. Fried, J. R.; Molecular simulation of gas and vapour transport in highly permeable polymers. *Mater. Sci. Membr. Gas Vapour Sep.* **2006,** 95–136.
122. El Sheikh, A.; Ajeeli, A.; and Abu-Taieh, E.; Simulation and Modeling: Current Technologies and Applications. IGI Publishing; **2007**.
123. McDonald, I.; NpT-ensemble monte carlo calculations for binary liquid mixtures. *Mole. Phys.* **2002,** *100(1),* 95–105.
124. Vacatello, M.; et al. A computer model of molecular arrangement in a n-paraffinic liquid. *J. Chem. Phys.* **1980,** *73(1),* 548–552.
125. Furukawa, S.-I.; and Nitta, T.; Non-equilibrium molecular dynamics simulation studies on gas permeation across carbon membranes with different pore shape composed of micro-graphite crystallites. *J. Membr. Sci.* **2000,** *178(1),* 107–119.

INDEX